Restrictions on International Trade in Steel

Restrictions on International Trade in Steel

Craig R. MacPhee
University of Nebraska,
Lincoln

Lexington Books
D.C. Heath and Company
Lexington, Massachusetts
Toronto London

Library of Congress Cataloging in Publication Data

MacPhee, Craig Robert, 1944
Restrictions on international trade in steel.

Includes bibliographical references.
1. Steel industry and trade. 2. Tariff on steel. I. Title.
HD9510.5.M2 382'.45'672 73-11650
ISBN 0-669-89516-4

Published simultaneously in Canada.
Printed in the United States of America.
International Standard Book Number: 0-669-89516-4
Library of Congress Catalog Card Number: 73-11650

To Paul, Heather, and Rob

Contents

	List of Figures	xi
	List of Tables	xiii
	Preface	xvii
	Acknowledgments	xix
Part I	*Measuring Influences on Trade*	1
Chapter 1	**An Overview of the Steel Trade**	3
	Trade Patterns	3
	Explaining Changes in Trade	4
	Trade Restrictions	17
	Defining Trade Restrictions	19
	The Plan for Analysis	19
	Notes	22
Chapter 2	**Tariffs and Nontariff Barriers**	27
	Tariffs	27
	Nontariff Barriers	36
	The Restrictive Effect of Trade Barriers	42
	Notes	48
Chapter 3	**Demand and Supply Elasticities**	51
	United States Import Demand Elasticity	51
	Variables in the Model	52
	The Import Demand Equation	57
	Domestic Demand and Supply Elasticities	58
	Foreign Import Demand Elasticities	60
	Export Supply Elasticities	63
	The Elasticity of Demand for Exports	67
	Notes	67
Chapter 4	**The Restrictiveness of Tariffs and Nontariff Barriers**	73
	Government Procurement Policies	73
	Quantitative Restrictions	78
	Subsidies	82
	Other Nontariff Barriers	86

	The Total Amount of Steel Restricted by Nontariff Barriers	87
	Tariff and Nontariff Restrictions	89
	The Impact of All Trade Restrictions	100
	Notes	101
Part II	*Nontariff Barriers to Trade*	103
Chapter 5	**Foreign Trade Policies**	105
	Import Licensing	105
	Foreign Exchange Restrictions	106
	Quotas	108
	Export Limitations	108
	Domestic Biased Procurement	109
	Subsidies	112
	Import Surcharges	122
	Antidumping Measures	123
	Notes	127
Chapter 6	**Administrative Practices**	135
	Import Marking Requirements	135
	Documentary Requirements	136
	Customs Classification	138
	Valuation	140
	Customs Fees and Deposits	144
	Other Customs Provisions	145
	Notes	146
Chapter 7	**Internal Policies and Regulations**	151
	Internal Economic Policies	151
	Other Internal Policies	160
	Notes	162
Chapter 8	**Restrictive Business Practices**	165
	Steel Cartels	165
	Restrictive Practices	166
	Exclusive Supply Agreements	167
	Freight Rate Discrimination	168
	Notes	176

Appendix	**Statistical Tables**	179
	Index	193
	About the Author	197

List of Figures

1-1	World Steel Production and Trade	5
1-2	Capacity Utilization and Export Prices	11
2-1	Nontariff Barriers Generally Imposed on All Imports	39
2-2	Nontariff Barriers Selectively Imposed on Some Imports	40
2-3	Nontariff Barriers Facing Steel Imports	41
5-1	The Monopoly Effect of an Import License	107
5-2	The Japanese Steel Export Market	115

List of Tables

1-1	World Steel Production	6
1-2	International Trade in Steel: Selected Countries	6
1-3	Dollar Cost Differentials per Ton of Steel	9
1-4	Growth of Output per Man-Hour in Selected Manufacturing Industries	12
1-5	Estimated R & D Expenditures as a Percentage of Sales (1972)	13
1-6	Percentage of 1970 Production by Continuous Casting	14
1-7	Nontariff Barriers	20
1-8	Nontariff Barriers to Trade in Manufactures	20
1-9	Nontariff Barriers	21
1-10	Nontariff Barriers to Trade	22
2-1	Steel Tariffs	28
2-2	1964 Steel Tariffs: Selected Countries and Products	29
2-3	1962 Nominal and Effective Tariff Rates on Steel	33
2-4	1962 Ranks of Steel Tariffs Among Those of Thirty-Six Industries	34
2-5	Selected U.S. Tariff Rates and Rates of Effective Protection	36
2-6	United States Nominal and Effective Tariff Rates on Steel	37
2-7	Quantified Items Creating Nontariff Barriers to Steel	43
3-1	Elasticities of Import Demand for Steel with Respect to Price	61
3-2	The Price Elasticity of the Supply of Steel Exports from Each of the Major Steel-trading Countries	64
3-3	The Price Elasticity of the Supply of Steel Exports to Each of the Major Steel-trading Countries	65
3-4	The Price Elasticity of Demand for the Steel Exports of Some Major Steel Traders	66
4-1	Steel Consumption, Imports, and Price Preferences of Governments in the Major Steel-trading Countries	75

4-2	Steel Imports Excluded by Government Purchasing Policies	76
4-3	Value of 1968 Imports of Iron and Steel Mill Products	80
4-4	Steel Imports Excluded by the French Quota	81
4-5	Steel Imports Excluded by the 1969 "Voluntary" Export Limit	81
4-6	Nontariff Barrier Rates Created by Domestic Subsidies	83
4-7	Steel Imports Excluded by Domestic Subsidies	84
4-8	Nontariff Barrier Rates Created by Export Subsidies	87
4-9	Steel Imports Excluded by Export Subsidies	88
4-10	Rates of Other Nontariff Barriers	89
4-11	Steel Imports Excluded by Other Nontariff Barriers	90
4-12	Steel Imports Excluded by All Nontariff Barriers	92
4-13	The Distribution of Nontariff Barriers	94
4-14	Steel Excluded by Quantified Nontariff Barriers	94
4-15	Steel Imports Excluded by Tariffs	96
4-16	Steel Imports Excluded by Tariffs and Nontariff Barriers	98
5-1	Price Preference Given Domestic Goods by Purchasing Governments	112
5-2	Direct and Indirect Taxes as a Percentage of Total Taxation	117
5-3	Direct and Indirect Taxes as a Percentage of GNP	117
5-4	Profits Tax as a Percentage of Corporate Profits	117
5-5	Export Subsidization Due to Border Tax Rebates	123
5-6	United States Antidumping Cases	127
6-1	United States Customs Court Classification Protest Cases	141
6-2	United States Customs Court Appeals for Reappraisement of Customs Valuation	143
6-3	United States Court of Customs Appeals: Average Number of Months between Filing and Decision	143
6-4	Days Allowed for Customs Entry after Arrival of Imports	145
7-1	First-Year Depreciation Allowances	153

7-2	Investment Incentives	154
7-3	Nontariff Barriers Created by Coking Coal Subsidies	156
7-4	Border Taxes on Imports	157
8-1	Conference Ocean Freight Rates on United States Exports to Major Steel-trading Countries	171
8-2	Conference Ocean Freight Rates on United States Imports from Major Steel-trading Countries	172
8-3	Conference Ocean Freight Rates on United States Exports to Major Steel-trading Countries	172
8-4	Conference Ocean Freight Rates on United States Imports from Major Steel-trading Countries	173
8-5	Conference Ocean Freight Rate Differentials	173
8-6	The Share of Steel Products in the Total Steel Trade of the Major Steel-trading Countries	174
8-7	Conference Ocean Freight Rate Differentials Weighted by the Share of Each Steel Product in Total Steel Exports	175
8-8	United States Freight Rate Differential Calculation	175
A-1	United States Imports of Iron and Steel Mill Products	179
A-2	Steel Price Indexes	181
A-3	Index of United States Gross National Product in Constant Dollars	183
A-4	United States Crude Steel Capacity	184
A-5	United States Crude Steel Capacity Estimates	184
A-6	Capacity Utilization in the United States Steel Industry	185
A-7	Collective Bargaining in the United States Steel Industry	187
A-8	Quarterly Periods of Steel Buying Affected by Steel Labor Negotiations	188
A-9	Man-Days Lost Due to Work Stoppages in the Water Transportation Industry	190
A-10	The Share of Imports in the Domestic and International Market	190
A-11	The Share of Exports in Domestic and International Markets	191

Preface

Steel, as the most basic of all manufactured goods, symbolizes industrialization. One strains to recall a commodity whose production or distribution does not utilize the metal. Small wonder, then, that steel stands out as an important target of economic policy. Steelmaking has been nationalized in some countries, subsidized, protected, and regulated in others. Whether success or failure best characterizes these policies remains the subject of considerable controversy, but it is my hope that this study contributes to resolution of the debate by focusing on the international repercussions of many policies.

Part I begins with a consideration of recent changes in the steel trade and evaluates a number of hypotheses intended to explain them. This leads naturally to an intensive analysis of one little explored but much talked about factor influencing trade, namely distortions produced by government policies. I also construct a partial equilibrium model in Part I to estimate the effects of these policies. While Part I summarizes the various tariffs and other trade barriers erected by the main steel trading countries, details on nontariff restraints have been left to Part II.

The work began with my doctoral dissertation on nontariff barriers, completed at Michigan State University in 1970. Since then, it involved review of other influences on trade flows, as well as measurement of the effective and nominal rates of protection accorded by tariffs.

I collected most of the original data on nontariff barriers appearing in Part II prior to 1970, but it has been possible to update some of this information, at least for the United States. Nevertheless, one may question the reliability of even the most recent published information because governments can make unannounced changes in these restrictions quite easily. The lack of international agreements or legislative statutes and the prevalence of administrative policies facilitate frequent and covert adjustments. Most foreign governments (unlike the United States government) prefer not to provide information on many of their restrictions, a practice which creates more uncertainty for importers and denies raw material for empirical research. The few official documents that do exist can be obtained only with considerable trouble. Consequently, we must depend on the sometimes subjective reports of private individuals, trade associations, international organizations, and journalistic sources.

Questionnaire responses from steel importers, exporters, producers, and consumers also provided information on trade restrictions that could not be obtained elsewhere. Respondents in some cases added to lists of nontariff barriers, estimated their impact, and ranked barriers according to their restrictiveness, but they gave no indication of recent changes in these restrictions, nor did they indicate any discriminatory treatment of different types of steel. The rankings (reported in Part II) were often inconsistent with other information on the height of nontariff barriers, and the rate of response (13 out of 300) was low. Hence, I could not use the ranks as an aid in quantifying barriers with otherwise unmeasurable heights. I rely instead on documented quantitative data, and estimates of restrictiveness in Part I are confined to measurable trade restrictions.

Despite these limitations, I trust that the results will not only prove to be of value to those interested in steel, but that the methods employed here can be used to evaluate trade restrictions on other products as well.

Acknowledgments

Many individuals and institutions deserve thanks for their help, but, of course, should not be held responsible for the results. Computer facilities, clerical assistance, and/or financial support were provided by the Institute for International Business and Economic Development Studies at Michigan State University, and the Department of Economics and Bureau of Business Research at the University of Nebraska, Lincoln. Walter Adams kindled my initial interest in the steel industry. Mordechai Kreinin advised me on those parts of the book that draw on my dissertation, written under his supervision. Jeffrey Roth and Duane Hackman aided with computer programming. Portions of the manuscript were reviewed by Robert Allen, Rodney Peterson, Harish Gupta, Bruce Allen, Harry Brainard, and Joel Dirlam. Rodney Peterson, Kenneth Gagala, Lawrence Pinson, John Tillman, a number of United States Department of State and Department of Commerce officials, and foreign as well as domestic steel companies were the sources for data that were often otherwise unobtainable. My writing had the benefit of reviews and encouragement from my parents, Craig and Dorothy MacPhee. Jan Mowrey and Sylvia Hanner typed the final draft of my manuscript.

Throughout this effort, my wife, Kathleen, has been at my side —sometimes consoling or inspiring me, but more often than not—working as statistician, typist, editor, keypunch operator, librarian, consultant, collater, and logistics expert. Somehow, she also has found time to pursue her own interests and to bear and raise the three little people to whom I lovingly dedicate this book. For all these things she has my unending gratitude and respect.

Part I
Measuring Influences on Trade

1 An Overview of the Steel Trade

The bulk of what follows contains a comparative appraisal of steel-trade distortions produced by tariffs and nontariff barriers. The rationale for such a study, however, lies deeply imbedded in controversies over the causes and consequences of shifting steel-trade patterns. Thus an examination of the empirical evidence, and of the hypotheses intended to explain it, provides an approporiate starting point.

Trade Patterns

As Figure 1-1 illustrates for the years in which statistics are available, total world exports of steel more than kept pace with production. The twentieth century to date witnessed a twentyfold increase in steel output with post-World War II expansion accounting for almost 80 percent of the increase. However, Table 1-1 shows that the output increases have not been uniformly distributed. Japan's twentyfold expansion contrasts markedly with the meager growth of American steel output.

As for trade, the First and Second World Wars as well as intervening years of international financial crises and protectionism disrupted the exchange of steel, but since then exports rose 1000 percent. This climb just paced the increase in total world exports, and in that sense it remains unremarkable. Nevertheless, the fact that the trade flow in ferrous metals changed course as it grew is noteworthy.

Five years after the end of World War II, the United States accounted for almost half of world steel output and one-sixth of world steel exports. As renewed industrialization progressed in the war-ravaged countries of Europe and Japan, and in Canada as well, their share of total world production rose dramatically and the share of their production entering export markets increased (e.g., by two-thirds in Japan and by 100 percent in Canada between 1955 and 1964).[1] Table 1-2 indicates that Europe, Canada, and Japan multiplied their exports many times over between 1950 and 1970.

The new export orientation of foreign steel producers quickly reflected itself in American imports, which rose from approximately 1 million tons in 1954 to a high of 18.3 million tons in 1971, according to United States Department of Commerce figures in Table A-1 of the Appendix. Meanwhile, United States steel exports did not share in the phenomenal growth of total world trade: they exceeded 4 million tons in only four of the last twenty years. Reflecting the poor export performance of American producers during the decade spanning the reversal of the United States steel-trade balance, imports as a percentage of Canadian, Japanese, and British consumption shrunk while the import shares of the original six Common Market members registered only small increases. The ten countries referred to here account for about 80 percent of production and 90 percent of exports in the free world. Although production and trade increasingly involve other countries, much insight can be gained through concentrating on the major steel traders.

Explaining Changes in Trade

Naturally, the transformation of a steel-trade surplus into a deficit attracts the attention of government officials concerned with balance-of-payments problems, unemployment, and the national security; steel firms and labor organizations respectively charged with augmenting the earnings of stockholders and workers; and, of course, the curious economist. Not surprisingly, the explanations of the trade reversal emanating from this melange vary in quality and content. Some emphasize such comparative advantage considerations as relative factor costs and technological progress, while others blame the development of excess capacity abroad, the oligopolistic behavior of United States firms, the effects of American labor disputes, or the discriminatory impact of foreign restrictions on imports from the United States.

Labor Costs

Different estimates exist for the differential materials, labor, and capital prices paid by the major steel producers, but most give little support to the argument that American steel cannot be cost competitive with the foreign product.

Payments to production workers, including all fringe benefits provided by the employer, comprise 42 percent of of steel production costs in the United States.[2] For 1966, Thorn found hourly wage costs of production workers in the United States ($4.63) exceeded those of other major producers by two to four times.[3] And American producers are quick to point out that their competitive disadvantage in price results from "sharply higher labor costs in this nation."[4] However, some countries such as Japan and France employ relatively large numbers of nonproduction workers in the steel labor force, and hourly wages in the foreign mills rose at 50 to 700

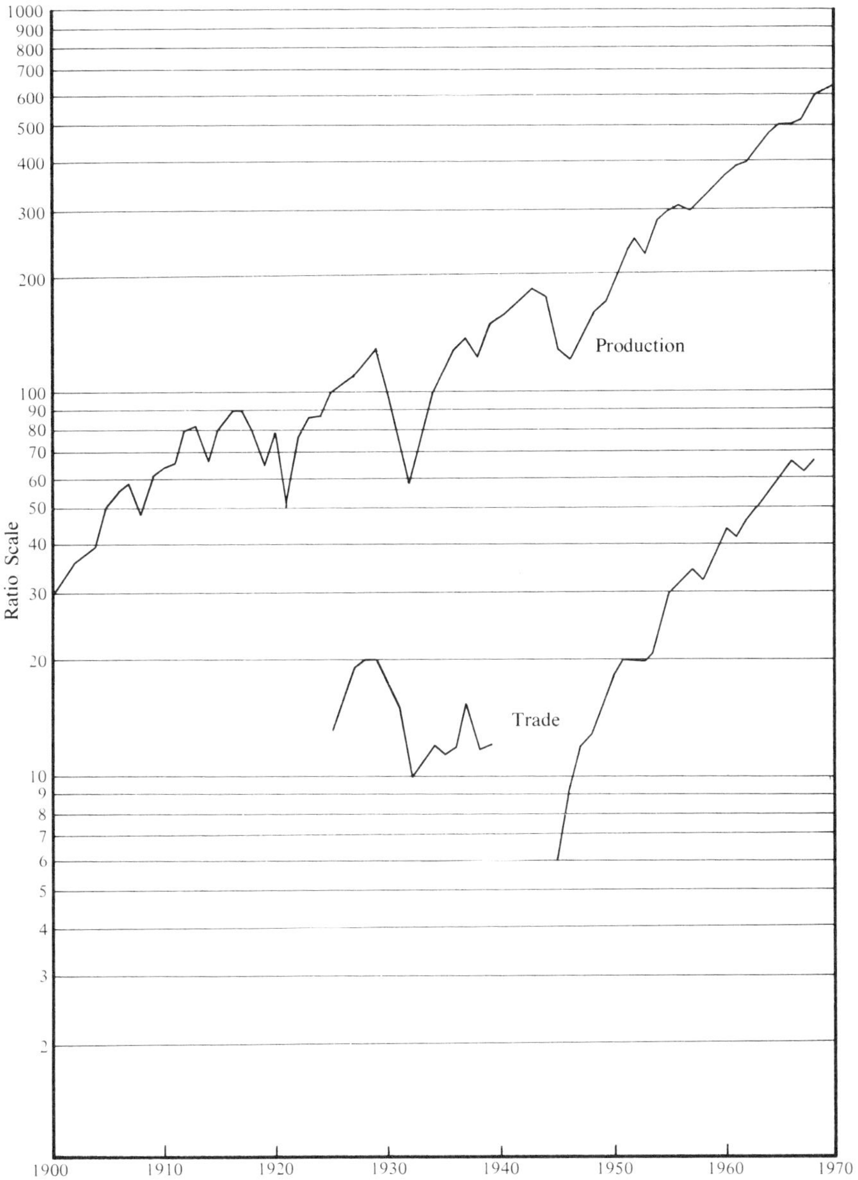

Source: American Iron and Steel Institute, *The Steel Import Problem*, p. 16.

Figure 1-1 World Steel Production and Trade (Millions of Tons)

Table 1-1
World Steel Production (Millions of Net Tons of Ingots)

	1940	*1950*	*1960*	*1970*
Canada	NA	NA	6	11
Japan	7	5	24	93
Western Europe	50	58	120	154
United States	67	97	99	119
USSR and Eastern Europe	25	39	95	155
All Others	6	9	36	62
Total	155	208	380	594

Sources: American Iron and Steel Institute (AISI), *Charting Steel's Progress,* 1961, pp. 38, 26; and *Iron Age* (January 4, 1973), p. 113.

Table 1-2
International Trade in Steel:[a] Selected Countries (1,000s of Tons)

	1950		*1960*		*1970*	
	Imports	Exports	Imports	Exports	Imports	Exports
Belgium-Luxembourg	120	3364	634	7564	2103	11849
Canada	NA	165[b]	695	663[b]	1219	1320
France	117	3283	2982	5394	6860	6464
W. Germany	210	1747	3968	7849	8391	9139
Italy	626	121	1749	1115	4676	1229
Japan	5	546	186	2242	113	14285
Netherlands	1122	189	2153	1160	2523	2797
UK	477	2389	1224	3101	2089	3042
US	1115	2576	2459	2710	9432	5747

[a]Ingots, finished, and semifinished steel.

[b]Ferro alloys and rolling mill products.

Sources: Economic Commission for Europe, *World Trade in Steel and Steel Demand in Developing Countries.* (New York: United Nations, 1968), pp. 66, 89, 90; and Organization for Economic Cooperation and Development (OECD), *The Iron and Steel Industry in 1971 and Trends in 1972* (Paris, 1973), pp. 7, 74.

percent faster rates over the 1964-66 period. In addition, wage rates must by multiplied by output per manhour in order to arrive at labor costs per ton of steel produced, but strictly comparable productivity data do not exist.

For one thing, many different factors—better raw materials, improved plant management, optimum capacity utilization, increased capital intensity, new technology, as well as upgraded worker skills—contribute to increases in output per manhour. For another, aggregate manhour measures ignore the mix of the labor force with respect to full- or part-time workers, skilled or unskilled workers, and the arbitrary nature of the distinction drawn between production and nonproduction workers, or employees and independent contractors, who perform many tasks in Japanese steel mills. The quality of raw materials like coal and iron ore differ significantly among countries. Changes in product mix (which are made by United States Steel Corporation alone on the average of once a week) and quality can also limit the relevance of productivity measures.

Notwithstanding those qualifications, Thorn's 1966 data show American productivity (7.0 production manhours per ton) leading all major steel producers except the Netherlands (6.9 manhours per ton). However, foreign steel producers witnessed relatively rapid increases in the 1960s. According to Bureau of Labor Statistics (BLS) figures for all workers, manhours per ton of steel fell 72 percent in Japan, 40 percent in West Germany, but only 2 percent in the United States. These changes bring Japan (at 12.05 total manhours per ton), according to the American Iron and Steel Institute (AISI), to a productivity parity with the United States (at 11.63 total manhours per ton). The conclusion follows that even in 1966 foreign steel producers enjoyed an average absolute labor cost advantage at the mill ranging between 6 and 15 percent of the unit price of steel. And in Japan unit labor costs declined during 1965-70 while the Americans recorded a 22 percent increase. To illustrate the significance of these facts, at the same rates of productivity advance, labor costs per ton would remain permanently higher in the United States, even if Japan paid American wages. Conversely, if Japanese productivity increases fell to the United States rate and Japanese wages continued their rate of increase, Japan would enjoy a competitive advantage in this regard for the next ten years.

Material Costs

We find the domestic prices of other inputs difficult to estimate accurately because most of the major steel producers do not acquire their raw materials on open domestic markets.[5] For instance, owners of the Belgian steel industry also control most of the coal mines, with the result that Belgian mills produce 75 percent of their own coke requirements. In addition, the Belgian steel companies control sufficient iron ore deposits in Luxembourg and France to cover their requirements. The Italian state controls most of the country's crude steel, coal, and iron ore production and the consequent steel complex in effect constitutes a large vertical concentration from raw materials to final products. The steel industry in Luxembourg is owned by highly integrated Belgian and German steel companies, one of which, for example, controls one-fourth of the Saar region's steel output, some coal mines in Germany, iron ore mines in France, fabricating plants in Belgium and Germany, as well as 60 percent of Luxembourg's 1955 crude steel output. The dominant American firms are also completely integrated.

Although we cannot obtain realistic domestic market prices, import prices may provide a sufficiently close approximation of material costs. Japan, for example, makes 90 percent of its raw material outlays abroad. Belgium imports over 20 percent of its coke requirements as well as iron ore from the Lorraine region of France and from Sweden. Italy imports all of its coal for coke production, and over half of its iron ore and scrap. Luxembourg uses its own low-grade of iron ore, but must import a higher grade from Lorraine as well as all of its blast furnace coke. The Netherlands imports about 40 percent of its coal requirements and virtually all of its iron ore. In contrast, the United States relies on domestic mines for two-thirds of its iron ore and nearly all of its coal. Canada also employs predominantly its own resources.

The f.o.b. import prices of iron ore are continuing in a long-term decline in the United States and abroad.[6] However, c.i.f. price decreases are less pronounced because of increases in port charges, unloading costs, inland transport, and storage costs, despite a decline in shipping rates. Scrap prices in Europe, Japan, and the United States have also generally declined over the last ten years. Though they rose in early 1968 in response to a somewhat sharp upturn in demand, scrap prices decreased markedly later in the year. The quantity of coke used in making a ton of pig iron has fallen annually in recent years, and the concurrent trend in coke prices has been the same. Increased demand in all countries and limited supply contribute to increases in fuel prices facing all steel producers.

If we can assume that world trade equalizes f.o.b. import prices, that the import prices fairly represent the opportunity costs of raw materials to steel producers, and that (following Thorn) Europe and Japan import all their metallurgical coals from the United States, then the only differences in coal costs would stem from transport expenses. Ignoring differences in iron ore and other input costs, Thorn estimates that Japanese producers pay $3.60 to $4.65 more for coking coal per ton of steel than American mills, and the Europeans $2.37 to $2.96 more. This estimate disregards the costs of iron ore, scrap, limestone, and a variety of fuels and alloys—many of which are probably higher in Europe and Japan. Thus it comes as no surprise that an even larger American advantage appears in 1964 Japanese Iron and Steel Federation (JISF) data reported by Thorn: Japan uses $44.40 of raw materials per ton of steel, the United States $36.00, the United Kingdom $38.50, and West Germany $43.50. Material costs account for less than half of total revenue in the United States, whereas the Japanese industry spends about 60¢ of each sales dollar on raw materials.[7]

Capital Costs

Thorn uses estimates of $330, $133, and $230 as the cost of a net ton of integrated capacity in the United States, Japan, and Europe respectively. He assumes the same depreciation and interest rates for each steel trader, and arrives at a capital cost advantage for Japan of $10.00 per ton and for Europe of $5.00 per ton. The JISF quotes capital costs in Japan at $20.70 per ton of steel produced, United States $16.60, United Kingdom $11.40, and West Germany $15.60.[8]

Total Cost Differentials

In sum, f.o.b. cost data from either Thorn or the JISF (shown in Table 1-3) can explain the relatively poor export performance of American steel producers. Japan can produce a ton of ferrous metal anywhere from $23 to $51 cheaper than the United States, and Europeans can outdo Americans by $13 to $37. However, the enlarged share of imports in the United States market cannot be explained entirely with either Thorn's data or Weidenhammer's figures from a congressional report. After adding tariffs ($7 per ton) and ocean freight charges ($13.97 to $22.33 per ton), Thorn finds the Japanese at a $1.64 to $6.25 cost disadvantage in the American market, and the Europeans possessing as much as a $15.24 disadvantage. Weidenhammer finds the Japanese with a $16 advantage, while European costs were on a par with American.[9] Thus it becomes useful to look at

Table 1-3
Dollar Cost Differentials per Ton of Steel

	Thorn		*JISF*		
	Japan	EEC	Japan	UK	W. Germany
Labor	−17.63	−18.62 to − 10.72	−63.30	−30.00	−43.60
Capital	−10.00	−5.00	4.10	− 5.20	− 1.00
Raw materials	3.60 to 4.65	2.37 to 2.96	8.40	2.50	7.50
Total	−24.03 to 22.98	−21.25 to −12.76	−50.80	−32.70	−37.10

Note: Minus sign indicates amount below U.S. costs.

Source: Based on figures reported by R. Thorn, "Steel Imports, Labor Productivity, and Cost Competitiveness," *Western Economic Journal* (December 1968), p. 379.

alternative as well as supplementary explanations of the steel-trade reversal.

Capacity

The United States steel industry contends that American and foreign government assistance led to the development of excess steel capacity overseas, and that as a result foreign producers dump steel in the United States.[10] Evidence presented in Chapter 7 confirms the role that foreign governments play in subsidizing the construction of plant and equipment. It is also true that the United States subsidized metal works abroad to the extent of $2,166 million in low-cost loans and grants. However, only one-fifth of this help went to the major steel-trading countries.[11]

World capacity and production data (excluding the United States) verify the existence of excess capacity abroad. Underutilization of capacity cannot assume the status of a recent problem, of course, since it plagued the Japanese steel industry as early as the 1920s. Nor is the current problem severe. The 1955-65 period spanning the reversal in the American trade balance witnessed a decline in utilization of facilities from 97 percent to 90 percent worldwide. The Japanese did no worse from 100 percent to 90 percent), although the EEC's production/capacity ratios (from 96 percent to 84 percent) more closely approximate American performance (from 93 percent to 80 percent).[12]

Casual inspection of the data for two years, however, hides the impact of short-term fluctuations in capacity utilization abroad (*C*) on the level of United States imports (*M*). The following regression equation offers a systematic test of this influence for 1954-68, standard errors appearing in parentheses.

$$M = \underset{(4.851)}{10.303} - \underset{(4.042)}{8.333}\,C \qquad R^2 = 0.278$$

The results indicate that the supply of exports to the United States is significantly correlated with excess capacity abroad, although the percentage of explained variance is small. A look at prices provides an important causal link. As Figure 1-2 illustrates, European steel export prices respond to excess capacity, and American buyers wisely purchase the cheaper metal.

Technology

A more fundamental explanation of the deteriorating competitive ability of

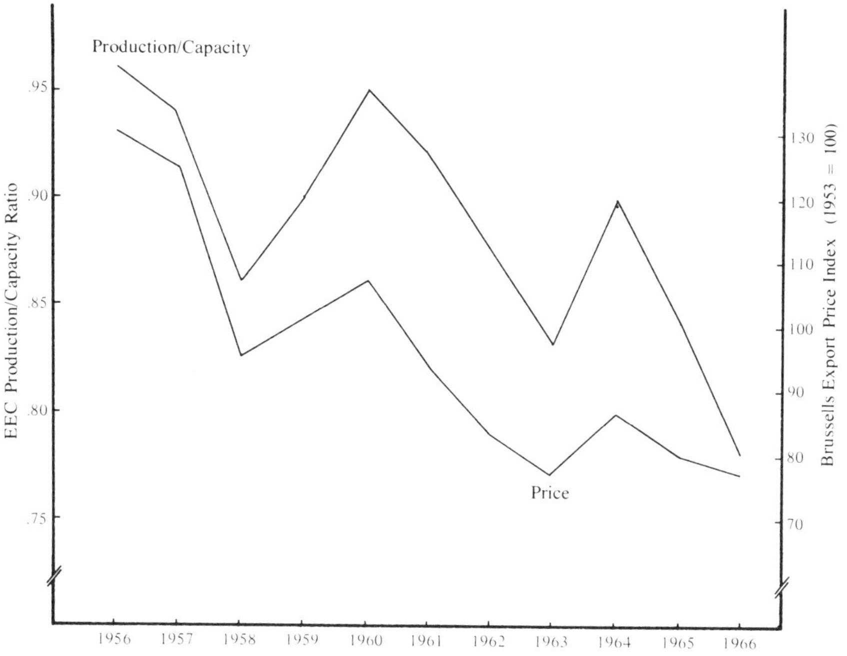

Source: American Iron and Steel Institute, *The Steel Import Problem,*, pp. 59, 65.

Figure 1-2 Capacity Utilization and Export Prices

United States steel—one that underlies differences observed in production costs—cites the technological lethargy of American producers. What position does the domestic steel industry hold in the technological race with steelmakers abroad? E. M. Gott, president of the AISI, replies, "We're out in front."[13] But the evidence does not sustain that claim. Not only have foreign producers increased labor productivity faster over the last decade (a fact we observed earlier in the chapter), but even in the United States, steel ranks behind most other industries in its rate of growth in output per manhour (as illustrated in Table 1-4).

In view of these figures, it is not surprising that Mansfield's detailed study of actual innovations (weighted by their importance) enabled him to conclude that the largest four firms in the less concentrated oil and coal industries carried out a larger number of innovations proportional to their market share than did those in the American steel industry.[14] Nor does the domestic steel industry appear to be exerting much effort toward overcoming the technology gap. Table 1-5 shows planned research and development expenditures as a percentage of sales for steel comparing unfavorably with similar budgets for fifteen major industries and an all-manufacturing average.

"Big Steel" has been roundly criticized for its failure to rapidly develop

Table 1-4
Growth of Output per Man-Hour[a] in Selected Manufacturing Industries (Average Annual Rates of Growth)

Industry	*1960-71*
Radio and television receiving sets	6.9
Hosiery	6.7
Malt liquors	6.4
Petroleum refining	5.5
Man-made fibers	5.5
Aluminum rolling and drawing	5.1
Bottled and canned soft drinks	5.1
Major household appliances	4.7
Concrete products	4.6
Cement, hydraulic	4.4
Paper, paperboard, and pulp mills	4.2
Sugar	4.2
Flour and other grain mill products	4.1
Cigars	3.9
Tires and inner tubes	3.7
Corrugated and solid fiber boxes	3.5
Motor vehicles and equipment	3.0
Candy and other confectionary products	3.0
Glass containers	2.7
Gray iron foundries	2.6
Primary aluminum	2.5
Canning and preserving	2.3
Tobacco products— total	2.2
Steel	2.2
Cigarettes, chewing and smoking tobacco	1.3
Primary copper, lead, and zinc	1.2
Footwear	0.3

[a]All employees.

Source: Bureau of Labor Statistics, "Indexes of Output Per Man Hour: Selected Industries, *Bulletin 1758* (Washington, D.C.: United States Government Printing Office, 1972), p. 3.

or adopt process innovations. Adams and Dirlam[15] note that none of the major changes in steelmaking—the Bessemer, Siemens-Martin (open hearth), or basic oxygen processes—were American inventions. Oxygen converters, in particular, were first commercially employed in a small Austrian steel plant (1952), first used on the North American continent by a Canadian plant, and first installed in the United States by McLouth Steel (1954)—a company possessing only 1 percent of the United States ingot capacity. Despite cost savings of over $5 per ton, oxygen furnaces in the United States accounted for only 25 percent of 1964 capacity, compared with 63 percent in Japan, over 70 percent in the Netherlands, and an average of 24 percent for the rest of Europe (where the process could not readily be adapted to most European ores). By 1969, 73 percent of Japan's capacity consisted of oxygen furnaces, as did 69 percent of Austria's, 46

Table 1-5
Estimated R &D Expenditures as a Percentage of Sales (1972)

Aerospace	20.88%
Electrical machinery and communications	7.72
Professional and scientific instruments	6.27
Chemicals	3.41
Machinery	2.96
Autos, trucks, and parts and other transp. equipment	2.05
Petroleum products	1.73
Rubber products	1.43
Nonferrous metals	.79
Paper	.47
Fabricated metals and ordinance	.40
Steel	.26
Textile mill products and apparel	.26
Food and beverages	.20
Other manufacturing	.14
All manufacturing	2.32

Source: McGraw-Hill as reported in D. Seifer, "The Case for Restricting Steel Imports," *Pittsburgh Business Review* (September-October 1972), p. 4

percent of West Germany's, and 46 percent of the Dutch, as compared to 43 percent of the American's.[16]

Ault[17] finds the United States lagging with respect to another major innovation: the continuous casting of semifinished forms directly from molten steel. Experimentation with this new process began as early as 1950 in West Germany, and by the early 1960s most major world producers, with the exception of big American firms (until 1967), installed continuous casting machines. As Table 1-6 delineates for 1970, American firms produce less steel by continuous casting than many other steel producers. Although estimated cost-savings range from $5 to $10 per ton, United States firms preferred to let others take the lead, allegedly because of confusion over patent rights, the height of the machinery, quality variations, and underestimates of continuous casting's potential in large-scale operations.

How have these developments affected world trade patterns? Ault's statistical tests support the contention that differences in the employment of the basic oxygen process and continuous casting offer significant explanations for changing trade patterns. Moreover, he reasons that countries adopting the new processes quickly and becoming the first to enjoy internal and external economies of scale, may continue to dominate world markets long after diffusion of the new technology to other steel producers.

Table 1-6
Percentage of 1970 Production by Continuous Casting (Selected Countries)

Austria	7.9	Italy	4.1
Canada	11.3	Portugal	4.6
Finland	69.0	Spain	13.4
France	0.8	U.K.	1.7
W. Germany	8.2	U.S.	[a]
Greece	18.2		

[a]Less than 1 percent.

Source: Based on figures reported by D. Ault, "The Continued Deterioration of the Competitive Ability of the U.S. Steel Industry: The Development of Continuous Casting," *Western Economic Journal* (March 1973), p. 90.

Imperfect Competition

The absence of competition in the American steel industry furnishes another explanation for the shift in direction characterizing the steel trade. Imperfectly competitive markets tend to perform inefficiently, exhibit high costs, and become technologically backward. Moreover, as suggested by Adams and Dirlam, and also by Thorn, the United States steel oligopolists do not follow the flexible pricing policies of their foreign competitors either at home or abroad. Increased American imports of steel wire caused small, nonintegrated fabricators to respond with lower prices and to go abroad for lower-priced supplies of their basic raw material. However, their large, vertically integrated rivals often "preferred to abandon the field to imports altogether rather than to reduce their product prices."[18] In export markets, the Iron and Steel Committee of the Organization for European Economic Cooperation observed that the United States companies are less willing than European and Japanese firms to make "sacrifices" in order to expand their share. Both the United States and the United Kingdom appear more reluctant to offer substantial price cuts to overseas customers.[19] Indeed, American steel's record of constant price escalation even looks bad next to the performance of other American firms. From 1947 to 1951, prices of basic steel products rose at the average rate of 9 percent a year, twice the average increase of all wholesale prices. The wholesale price index fell 0.9 percent annually from 1951 to 1955, but the price index for steel rose an average of 4.8 percent per year. Between 1955 and 1958, steel prices increased 7.1 percent annually, or almost three times as fast as wholesale prices generally. According to the Council of Economic Advisors, "No other major sector shows a similar record."[20]

No doubt exists about the lack of competition in the domestic industry. The *Annual Survey of Manufactures* for 1966 lists a four-firm concentration ratio of 49 percent for steel. The companies integrate vertically from iron ore and coal mines to fabricating plants. According to Bain's calculations,[21] entry into the steel industry requires more capital ($265 to $665 million for one efficient plant) than entry into the cement, petroleum, gypsum, rubber, or automobile industries.

An American plant must supply as much as 0.7 to 2.5 percent of industry output in order to take full advantage of economies of scale, and a steel firm of optimal size can control as much as 20 percent (or as little as 2 percent) of the national market. Bain also estimates that United States steel firms can raise their prices from 5 to 8 percent above average costs without attracting new entrants. Small wonder, then, that the same eight firms dominated the domestic industry for the last forty years.

This explanation for poor American performance loses some of its weight, however, when we look at the structure of industries abroad.[22] Foreign steel mills are generally smaller than those of the United States, ranging from less than one-tenth the size in Canada and Italy, to almost one-half in France; the United Kingdom plants are about one-third the United States size, while Japan's are one-fifth. Nevertheless, we can perceive structural similarities among the major steel industries by comparing concentration ratios for the period between 1958 and 1960. They show that in both Japan and the United States, the two largest firms accounted for about 40 percent of output. The same market share was claimed by the largest four firms in Britain, the largest three in France, and the largest one in Canada and Italy. In all the major steel-trading countries, concentration is greater in the production of primary forms than in advanced fabrication. Of course, these comparisons between the United States and other countries may differ now because of declining concentration in the United States industry and increasing merger activity in Europe and Japan. The Common Market Commission has said it would favor combining the EEC's 100 steel companies into ten major groups.[23]

Another qualification of firm concentration ratios stems from the fact that in France, Belgium-Luxembourg, and Germany, steel firms are controlled by interlocking groups composed of large producers and governments.[24] The groups may be holding companies, interlocking directorates, or members may be parties to long-term contracts. Six French groups controlled 76 percent of French steel production in 1958; the Belgian Societe Generale controls 36 percent of Belgian capacity and has an interest in 87 percent of Luxembourg and 23 percent of Saar capacity. The Italian state controls 55 percent of crude steel production in Italy. In the Netherlands, Roogovens produces three-quarters of the crude steel and controls 100 percent of the finished steel production, plus one of the largest steel producers in Germany. Despite the Allied Forces' deconcentration program during post-World War II occupation, eight groups controlled 99 percent of Ruhr, and 85 percent of German, production in 1959, as compared with the six groups which controlled about the same shares in 1930.

Of course, the pressures of international trade force even monopolists and socialists to act like competitors. As a glance at Tables A-10 and A-11 of the Appendix shows, all of the major steel traders except the United States export large shares of their output. Eighty-three percent of Belgium-Luxembourg production enters foreign markets; even Canada's export/production ratio of 7 percent is nine times greater than that of the United States. Imports also discipline foreign markets, with Belgium-Luxembourg importing 38 percent of their domestic consumption, the Netherlands 67 percent, and France 25 percent. Japan's import/ consumption ratio of 0.7 percent constitutes an important exception, but it is probably fair to conclude that foreign involvement in international trade provides a spur to competition abroad that the United States misses.

American Labor Disputes

The United States steel industry claims that "labor disputes have given foreign steel producers repeated opportunities to 'get their feet in the door' and afforded domestic steel users strong encouragement to try foreign sources of supply."[25] A look at the activity in the steel market before, during, and after actual or threatened strikes supports this argument. Before walkouts are scheduled, steel users increase their purchases from domestic and foreign suppliers to build up inventories for an anticipated strike. During a strike, customers may go abroad to satisfy their steel requirements. After a strike, the strain of high demand to replenish inventories taxes the capacity of the domestic industry. Resulting delays and price increases may prompt many customers to buy overseas.

Refuting the AISI argument, a University of Michigan researcher found little permanent loss of markets to foreign steel because of strikes.[26] However, after surveying the labor disputes and import trends of the last decade, a congressional study concluded that hedge buying has been a major factor in the import picture.[27] We can obtain another validation of the hypothesis from Rippe, whose econometric model to explain American import shares contained a statistically significant dummy variable representing the impact of labor negotiations in the second quarter of 1965.[28] Finally, the import demand function estimated in Chapter 3 also includes a statistically significant dummy variable to account for hedge-buying periods affected by steel labor disputes and negotiations between 1954 and 1968. Thus labor problems appear to explain some of the past fluctuations in American steel imports. Their current explanatory power, however, loses strength because secret negotiations and preliminary no-strike pledges now prevail.

Trade Restrictions

Turning at last to the major subject of this volume, we find tariffs and nontariff barriers cited as a strong force in the steel trade. The president of the United States has pronounced that our trade balance in general suffers because we compete with one arm tied behind our back. Picking up on that theme, an AISI president stated that "American steel is denied competitive access" to large foreign markets because of tariffs and nontariff barriers. "Our steel industry . . . has been attempting to compete in world markets against steel industries of other nations whose governments provide them with direct and indirect subsidies for exports as well as direct and indirect barriers to imports."[29]

Raising Barriers to Trade

The continued interest in trade restrictions would seem inappropriate in view of steadily declining tariff rates were it not for the steady substitution of nontariff barriers for lower duties.[30] In part, the proliferation of nontariff barriers stems from the popularization of protectionism, and in the United States steel now serves as the rallying point for those advocating protectionist foreign trade policies.

Ever since the United States market became the object of inroads by foreign steel, American producers began filing antidumping charges, and appealing for relief through escape-clause actions, import surcharges, higher tariffs, and quotas. (These are detailed in Part II.) Political pressure on our government led it to coerce foreign steel trade associations into "voluntarily" limiting their exports to the United States in 1969 to 14 million tons. Allowable import growth of 5 percent per year and lack of quantitative limits on particular steel products, however, prompted the domestic industry to press for an even more stringent pact. In response, the United States government negotiated a new quota with producers abroad that imposed limits on specific products and restricted import growth to only 2.5 percent per year.

Foreign producers agreed to the "voluntary" restrictions under the threat of congressional action. Specifically, the Foreign Trade and Investment Act of 1973, better known as the Burke-Hartke bill, proposes quotas for all imports not already covered by quantitative restrictions and a new Foreign Trade and Investment Commission (including industry and labor representatives) to set quotas, rule on escape-clause cases, and expedite antidumping actions. Big business and labor back the bill, and steel spokesmen have attacked an alternative White House measure, though it is only slightly less restrictive.

Under President Nixon's 1973 trade bill, he would have the authority to raise or lower tariffs, quotas, and temporary surcharges. But in testimony before the House Ways and Means Committee in May 1973, Stewart S. Cort, chairman of Bethlehem Steel and AISI president expressed disappointment with "the flabby safeguards and fair trade provisions of the Administration's trade bill."[31] According to the AISI, the steel industry would prefer an "automatic triggering mechanism" for relief from import competition, a provision halting further tariff reductions, and a legislative authorization for presidentially negotiated agreements limiting foreign exports to the United States. Very likely, the last recommendation came in response to the only blow struck for consumers amid this flurry of protectionism. For in January, 1973, United States District Court Justice Gerhard Gessel ruled in a Consumers Union suit that the president had no right to grant antitrust immunity to foreign producers entering into voluntary export agreements. The State Department, of course, is appealing the judgment.

Rationalizing Trade Barriers

The United States steel industry and its supporters justify protective measures on the grounds that imports threaten national security, balance-of-payments equilibrium, the domestic industry's ability to invest, and the jobs of thousands of steel workers. These arguments have been critically examined elsewhere, but we may profitably summarize the examination here.[32] In fact, steel firms have long experienced shortages of manpower, and minimized hiring of the hard-core unemployed. Moreover, appropriate monetary and fiscal policies can offer assurance of high employment rates despite imports. Profits to finance new capacity appear low partly because of the industry's failure to write down obsolete excess capacity. The balance of payments would actually be hurt more by measures that raise the cost of steel inputs to our major exporters and encourage foreign retaliation. Finally, the national security argument ignores changes in the nature of warfare, our dependence on foreign ores for alloys, and our current capacity to divert ouput from civilian to military uses.

Evaluating Nontariff Barriers

Despite the preoccupation of public and private policymakers with nontariff barriers like "voluntary" quotas and national security restrictions, H. G. Johnson still found them lying in the "areas of ignorance" in 1967. According to him, "while a great deal has been said about the importance of nontariff barriers to trade . . . the effects of these barriers remain to be quantified.''[33] What has been said about them does largely consist of general descriptions. Bidwell, de Haas, Grunzel, and more recently Massel and Kelly mainly present institutional surveys of the restraints without analyzing their effectiveness.[34]

Massel attributes this failure to analytical problems created by unmeasurable institutional factors surrounding the subtle workings of nontariff barriers. These include the covertness of many administrative practices, and the uncertainty of arbitrary changes in the regulation of imports. In addition, Kelly mentions the difficulties of expressing the degree of discrimination against foreign goods in numerical terms. However, these problems did not prevent Haight, Heuser, Viner, Whittlesey, and lately Grubel, Johnson, Aliber, Stein, and the Joint Economic Committee of Congress from analyzing the trade ramifications of quotas, various taxes, state trading, and discriminatory freight rates.[35] A recent Brookings Institution study by Robert E. Baldwin provides not only one of the most complete descriptions of the different types of barriers, but also goes on to analyze effective protection rates for many of them.[36] These analytical successes provide some encouragement to proceed with our own inquiry.

Defining Trade Restrictions

As traditional instruments of protection, tariffs are recognizable to most people as special taxes levied on commodities when they cross national boundaries. Although some countries levy export taxes, the most familiar tariffs are import duties.

Unlike tariffs, nontariff barriers are not even easily defined. Massel broadly refers to them as any import restriction besides tariffs. Kelly, on the other hand, confines the name to government restrictions on imports, excluding restrictive private practices and other barriers erected by language, geography, history, or culture. Even more abundant than definitions are lists of nontariff barriers. Tables 1-7 to 1-9 furnish us with a sample of the various items included and the alternative ways they are classified. In order to make our examination of nontariff barriers to steel comprehensive, we use a synthesis of these inventories (shown in Table 1-10) as a guide to this unexplored area.

The Plan for Analysis

Chapter 2 begins by summarizing information on tariffs and on the nontariff barriers examined in Part II. Where possible, we convert nontariff barriers

Table 1-7
Nontariff Barriers

Quotas and restrictive state-trading policies
Export subsidies and taxes
Discriminatory government and private procurement policies
Selective indirect taxes
Selective domestic subsidies and aids
Restrictive customs procedures
Antidumping regulations
Restrictive administrative and technical regulations
Restrictive business practices
Controls over foreign investment
Restrictive immigration policies
Selective monetary controls and discriminatory exchange-rate policies

Source: R. Baldwin, *Nontariff Distortions of International Trade*, Washington, D.C.: The Brookings Institution, 1970.

Table 1-8
Nontariff Barriers to Trade in Manufactures

Foreign Trade Policies
Licensing requirements
Quota restrictions
Negotiated export limitations
Foreign exchange restrictions
State trading
Procurement policies favoring domestic products
Antidumping and similar regulations
Subsidies to exports

Administrative Practices
Classification of goods for customs purposes
Documentary, marking, and packaging requirements
Incomplete or delayed publication of customs information

Internal Economic Policies Affecting Imports
Internal taxes for revenue purposes
Taxes applied to imports to compensate for indirect taxes borne by comparable domestic goods
Pricing policies and price control regulations
Restrictions on advertising of goods

Internal Health and Safety Regulations Affecting Imports
Sanitary regulations
Technical specification requirements
Regulations applied for national security reasons

Source: UNCTAD Secretariat, *Measures for Expansion of Markets in Developed Countries for the Exports of Manufactures and Semi-Manufactures of Developing Countries* (United Nations Conference on Trade and Development, E. Conf. 46/PC/20, May 6, 1963).

to a tariff-like ad valorem rate as a percentage of import price. Many important restrictions, however, are not amenable to such a transformation. As a result, we confine ourselves to estimating the restrictiveness of tariffs and quantifiable nontariff restraints.

We derive estimates of domestic, import, and export demand elasticities for the United States and nine foreign countries in Chapter 3. The

Table 1-9
Nontariff Barriers

Customs Law
Regulations governing the right to import (e.g., licensing)
Valuation and appraisement of imported goods
Classification of goods for customs purposes
Marking, labeling and packaging requirements
Documentary requirements (including consular invoices)
Measures to counteract disruptive marketing practices, e.g., antidumping and countervailing duties
Penalties (for example, fees charged for mistakes on documents)
Fees assessed at customs to cover cost of processing (handling) goods
Administrative exemptions (for example, administrative authority to permit duty-free entry of goods for certain purposes)
Treatment of samples and advertising material
Prohibited and restricted imports (embargoes and quotas)
Administration of customs law provisions (delay in processing goods, inadequate or delayed publication of customs information)

Other Legislation Specifically Applicable to Imports
Taxes (e.g., excise, turnover)
Restrictions imposed to protect individual industries (e.g., quotas)
Exchange controls: foreign exchange may be allocated only for imports for certain types of merchandise
Restrictions applied for national security reasons (other than under customs law)
State trading (or the operation of enterprises granted exclusive or special import privileges)
Sanitary regulations (other than under customs law)
Food, drug, cosmetic, and pharmaceutical regulations
Patent, trademark, and copyright regulations
Shipping and insurance regulations

Other Legislative and Administrative Trade Barriers
Government purchasing regulations and practices
Domestic price control regulations
Restrictions on the internal sale, distribution, and use of products
Screen quotas and other restrictions affecting motion picture film and TV program material
Specifications, standards, and safety requirements affecting such products as electrical equipment, machinery, and automobiles
Internal taxes that bear more heavily on United States goods than on domestic products (for example, automobile taxes in Europe based on horsepower rating)
Restrictions on advertising of goods
Restrictions on display of goods at trade fairs and exhibitions

Source: United States Department of Commerce as printed in Senate Finance Committee, *Steel Imports* pp. 48, 49.

same chapter also presents alternative estimates of steel supply elasticities for the United States and the rest of the world. We employ the elasticities to estimate the value of steel imports excluded by each country's trade restrictions in Chapter 4, which concludes Part I.

Table 1-10
Nontariff Barriers to Trade

Foreign Trade Policies
Licensing requirements
Quota restrictions
Negotiated export limitations
Foreign exchange restrictions
State trading
Domestic biased procurement
Antidumping regulations
Export subsidies
Shipping and insurance regulations
Import surcharges
Administrative Practices
Classifications of goods for customs
Documentary requirements
Marking requirements
Penalties for procedural errors
Prior deposit requirements
Import valuation procedure
Customs fees
Treatment of samples and advertising
Internal Policies and Regulations
Domestic subsidies
Direct taxes
Indirect taxes
Price controls
Restrictions on advertising
Patent, trademark, and copyright laws
Credit controls
Sanitary regulations
Technical specification requirements
National security restrictions
Private Practices
Freight rate discrimination
Exclusive supply agreements
Other steel cartel activity

Notes

1. American Iron and Steel Institute (AISI), *The Steel Import Problem*, New York, 1967, p. 17.

2. W. Hogan, *The 1970s: Critical Years for Steel* (Lexington, Mass.: D. C. Heath, 1972), p. 25.

3. Labor cost and productivity data in this section are from J. van B. Dresser, Jr., T. Hout, and W. Rapp, "Competitive Development of the Japanese Steel Industry," in J. Cohen, ed., *Pacific Partnership: U.S.-Japan Trade*, Lexington, Mass.: D.C.Heath, 1973; W. Hogan, *The 1970s*; and R. Thorn, "Steel Imports, Labor Productivity, and Cost Competitiveness," *Western Economic Journal* (December 1968), pp. 375-84.

4. S. Cort, Chairman of Bethlehem Steel and President of the AISI as quoted in *Iron Age* (May 31, 1973), p. 15.

5. For confirmation of facts in this and the following paragraphs, see L. Lister, *Europe's Coal and Steel Community: An Experiment in Economic Union*, (New York: Twentieth Century Fund, 1960), pp. 40, 42, 44, 45, 140, 141.

6. Information on import prices can be found in Organization for Economic Cooperation and Development (OECD), *The Iron and Steel Industry in 1967 and Trends in 1968* (Paris, 1968), pp. 46-54.

7. Data in this paragraph are from Thorn, "Steel Imports," van B. Dresser, et al., "Competitive Development," and AISI, *Charting Steel's Progress*, various years.

8. Thorn, "Steel Imports," pp. 379, 381.

9. Ibid.

10. AISI, *Steel Import Problem*, p. 1.

11. Senate Finance Committee, *Steel Imports* (Washington, D. C.: United States Government Printing Office, 1967), p. 32.

12. Economic Commission for Europe, *The European Steel Market*, New York: United Nations, various years; and OECD, *The Iron and Steel Industry*, various years.

13. E. Gott as quoted in AISI, *Steel Facts* (Fall 1972), p. 4.

14. E. Mansfield, "Innovation and Size of Firm," *Monopoly Power and Economic Performance* (New York: Norton, 1964), p. 62.

15. W. Adams and J. Dirlam, "Big Steel, Invention and Innovation," *Quarterly Journal of Economics* (May 1966), pp. 167-89.

16. 1964 data are from D. Ault, "The Determinants of World Steel Exports: An Emipirical Study," *Review of Economics and Statistics* (February 1972), p. 44; and 1969 data are from van B. Dresser et al., "Competitive Development," p. 214.

17. D. Ault, "The Continued Deterioration of the Competitive Ability of the U.S. Steel Industry: The Development ofContinuous Casting," *Western Economic Journal* (March 1973), pp. 89-97.

18. W. Adams and J. Dirlam, "Steel Imports and Vertical Oligopoly Power," *American Economic Review* (September 1964), pp. 626-55.

19. Organization for European Economic Cooperation, *The Iron and Steel Industry in Europe* (Paris 1960), p. 97.

20. Council of Economic Advisers, *Report to the President on Steel Prices* (Washington, D. C.: United States Government Printing Office, 1965).

21. J. Bain, "Economies of Scale, Concentration, and the Condition of Entry in Twenty Manufacturing Industries," *American Economic Review*, March 1954, pp. 23, 30, 36; and idem, *Barriers to New Competition: Their Character and Consequences in Manufacturing Industries*, (Cambridge, Mass.: Harvard University Press, 1956), pp. 169-72.

22. This data comes from J. Bain, *International Differences in Industrial Structure* (New Haven, Connecticut: Yale University Press, 1966), pp. 20, 36, 78, 82, 92, 97, 104.

23. Subcommittee on Antitrust and Monopoly, Committee on the Judiciary, United States Senate, 88th Cong., 2nd sess. *Concentration Ratios in Manufacturing Industries*, Pt. 1., 1965, (Washington, D. C.: United States Government Printing Office) p. 8; "The World Battle for Steel," *Business Week* (June 4, 1966), p. 63; "Business Around the Globe: Trouble for Steel," *Fortune* (May 1966), p. 64,; "Mergers: Japanese Fever," *Time* (May 24, 1968), p. 68; and "Ten Groups Will Be Right for The Community," *European Community* (March 1970), p. 29.

24. Lister, *Europe's Coal and Steel Community*, pp. 131-68.

25. AISI, *The Steel Import Problem*, p. 12.

26. C. Rehmus, "Collective Bargaining in the Basic Steel Industry," *Proceedings of the American Statistical Association*, September 7, 1962.

27. Senate Finance Committee, *Steel Imports*, p. 152.

28. R. Rippe, "Wages, Prices, and Imports in the American Steel Industry," *Review of Economics and Statistics* (February 1970), p. 42.

29. E. Gott, quoted in AISI, *Steel Facts* (Fall 1972), p. 3.

30. See A. Malabre, Jr., "U. S. Businessmen Say Nontariff Bars to Goods Pose Growing Problem," *The Wall Street Journal* (November 26, 1968), p. 16; "Nontariff Tricks," *Time* (January 12, 1968), p. 55; R. Lawrence, "Trade Problems Face U. S. in 1969," *European Community*, no. 119, (December 1968-January 1969), p. 7; J. Boner, "Nontariff Curbs Spread Despite Free Traders' Fight to Remove Them," *The Wall Street Journal* (March 3, 1969), p. 1.

31. S. Cort, as quoted in *Iron Age*, May 31, 1973, p. 15.

32. F. Molz, "The Political Economy of Steel Import Quotas," *Journal of Economic Issues* (June/September, 1970), pp. 60-76.

33. H. Johnson, *Economic Policies Toward Less Developed Countries*, (Washington, D. C.: The Brookings Institution, 1967), pp. 245, 246.

34. P. Bidwell, *The Invisible Tariff: A Study of the The Control of Imports into the United States* (New York: Council on Foreign Relations, 1939), p. 4; J. Grunzel, *Economic Protectionism* (Oxford: Clarendon Press, 1916), p. 188.; J. de Hass, *The Practice of Foreign Trade* (New York: McGraw-Hill, 1933), p. 358; M. Massel, *Non-Tariff Barriers as an Obstacle to World Trade*, Reprint no. 97 (Washington, D. C.: The Brookings Institution, 1965: and W. Kelly, Jr., "Nontariff Barriers," *Studies in Trade Liberalization* (Baltimore: The John Hopkins Press, 1967), p. 265.

35. H. Heuser, *Control of International Trade,* London: Routledge, 1939; F. A. Haight, *French Import Quotas*, London: King, 1935; J. Viner, *Trade Relations Between Free Market and Controlled Economies*, Geneva: League of Nations, 1943; H. Grubel and H. Johnson, "Nominal Tariff Rates and U.S. Valuation Practices," *Review of Economics and Statistics*, May 1967; Joint Economic Committee, *Discriminatory Ocean Freight Rates and the Balance of Payments*, 89th Cong., 1st sess., pt. 3, 1965, Washington, D.C.: United States Government Printing Office, p. 450; R. Aliber and H. Stein, "Price of United States Exports and the Mix of Direct and Indirect Taxes," *American Economic Review* (September 1963), pp. 703-10; and C. Whittlesey, "Excise Taxes as a Substitute for Tariffs," *American Economic Review* 37 (December 1937).

36. R. Baldwin, *Nontariff Distortions of International Trade*, Washington, D.C.: The Brookings Institution, 1970.

2 Tariffs and Nontariff Barriers

In the last chapter we examined changes and the probable causes of changes in steel-trade flows. We ended with reference to one little explored cause—trade restrictions. Before proceeding to estimate their impact, however, we must enumerate the relevant tariffs and nontariff barriers, as well as learn how to measure them. That is our intent in this chapter.

Tariffs

Unweighted averages of post-Kennedy Round nominal tariff rates on steel imports entering nine of the ten major steel-trading countries appear in Table 2-1. We can employ these duties to determine the value of steel excluded by nominal tariffs. But before embarking on such a task, we must note three qualifications: one relating to the base used in calculating tariff rates, a second to aggregation problems, and the third to the use of nominal rates.

The Rate Base

The major steel traders levy specific duties on some and ad valorem duties on other steel products. For consistency, we convert specific rates of dollars, francs, and so forth per ton to ad valorem rates expressed as a percentage of price. Care must be exercised in this transformation, and in making comparisons of ad valorem rates between countries because not all use the same price as a base.

In particular, the AISI claims that United States steel producers compete at a disadvantage because "the United States ad valorem tariffs are computed on the declared value of the product f.o.b. port of departure, while the practice elsewhere in the world is to compute them on landed value." The AISI estimates the difference in value to equal 15 percent on steel products, and maintains that "it is not clear whether GATT [General Agreement on Tariffs and Trade] negotiations in determining the value of tariff concessions took into consideration the United States method of valuation."[1] However, Kelly finds that import duties in c.i.f. countries run

Table 2-1
Steel Tariffs (Unweighted C.I.F. Rates)

Product	*Pre-Kennedy* Round	*Post-Kennedy Round*
Canada	0.067	NA
European Economic Community	0.099	0.072
Japan	0.146	0.073
United Kingdom	0.113	0.090
United States	0.099	0.087

Sources: Maxwell Stamp Associates, "The Free Trade Area Option: Opportunity for Britain," *The Atlantic Trade Study* (London 1968), p. 40; and Senate Finance Committee, *Steel Imports* (Washington, D.C.: United States Government Printing Office, 1962), p. 48; and J. Melvin and B. Wilkinson, *Effective Protection in the Canadian Economy* (Economic Council of Canada, 1968), p. 25.

only 10 percent higher on the average than the receipts from the same rates on identical products in countries using an f.o.b. system of valuation. He also argues that such differences in valuation systems do not constitute nontariff barriers since it is possible to compare duties by expressing them on a common base.[2] Moreover, according to William M. Roth, the president's special advisor on trade negotiations, the differences between c.i.f. and f.o.b. valuation have been taken into consideration in tariff negotiations.[3]

Unweighted Averages

Steel tariffs shown in Table 2-2 display considerable variance, yet we must aggregate tariff rates to make them manageable in our broad study of the steel trade. This raises the problem of how to arrive at a true mean. Unweighted tariff averages ignore the relative importance of various commodities in total trade. But, all weighting techniques are subject to some objections. Weighting by imports gives low values to very prohibitive restrictions; weighting by domestic production and consumption is improper because of the distorting effects of trade restraints on consumer choice and intercommodity differences in tradability. Even weighting restrictions by the total value of world trade will still be affected by intercountry similarities in the structure of tariffs or other trade barriers, although this is probably the most legitimate means.

To illustrate the different results obtained by alternative weighting schemes, the post-Kennedy Round tariff rates in Table 2-1 diverge sharply from rates weighted by import shares: 6.4 percent for Europe and 6.8 percent for the United States.[4] For the post-Dillon Round period, we find

Table 2-2
1964 Steel Tariffs: Selected Countries and Producte (Percentage of Value)

	United Kingdom			
	Ad valorem Rate	*Specific Rate*[a]	*European Economic Community*	*United States*
Pig iron	10.0	2.10.0	5.0- 7.0	0.4- 1.0
Ingots	10.0	3.05.0	7.0	8.5-10.5
Billets	10.0	4.00.0	8.0-10.0	8.5-10.5
Hot-rolled strip[b]	10.0	6.05.0	7.0- 9.0	8.0- 9.8
Universals	10.0	4.15.0	9.0	8.0- 9.8
Wire rod	10.0	4.10.0	8.0-10.0	2.7- 6.5
Sections	10.0	4.10.0	8.0-10.0	2.2- 7.5
Hot-rolled strip	10.0	4.10.0	10.0	6.0- 9.5
Cold-rolled strip	10.0	4.10.0	10.0-13.6	6.0- 9.5
Sheets under 3mm	10.0	6.05.0	9.0-10.0	8.0-10.0
Sheets over 3mm	10.0	4.15.0	9.0	8.0-10.0
Wire, drawn	25.0	20.00.0	9.0-13.6	2.0-13.0
High carbon, alloy steels	10.0	3.05.0 to 7.05.0	free or 4.8-14.5	2.2-16.5
Railway rails	10.0	4.05.0	10.0-16.6	1.0- 5.0
Cast-iron pipes	17.5	—	9.0-14.0	10.0
Welded, seamless tubes	17.5-20.0	—	8.0-15.5	1.6-12.0

[a]Alternative duty expressed in £ per ton.

[b]For further rolling.

Source: Economic Commission for Europe, *World Trade in Steel and Steel Demand in Developing Countries (New York: United Nations, 1968), p. 25.*

nominal tariff rates in the United States of 7.7 percent using the total exports of the major steel traders as weights. In contrast, weighting by the relative value of total United States shipments yields 7.0 percent as the duty applied to blast furnace and steel mill products.

There have been few weighting problems with nontariff barriers in this study because most of them are applied across the board on steel products. In any event, the unweighted average of steel duties can give us a basis for an up-to-date, if not precise, comparison of the relative restrictiveness of nominal tariff and nontariff barriers.

Effective Rates of Protection

The final limitation besetting our nominal tariff (or nontariff barrier) estimates stems from their failure to indicate the effective rates of these restrictions. Effective tariffs take account of barriers to imported material inputs and indicate the protection accorded to domestic value-added. We usually indicate the restrictive effects of tariffs or nontariff barriers by the difference between potential and actual steel imports, when the former refers to trade flows that would take place under *ceteris paribus* assumptions. Therefore, we assume that restraints on other goods (in particular steel inputs) stay constant.

However, across-the-board reductions in tariff or nontariff barriers yield different results. It should be readily apparent that lowering restrictions to raw materials and other intermediate products would expand the domestic supply of final products. As a result, the increase in imports of final goods attributable to elimination of trade restraints would shrink. Moreover, the size of this increment in imports will vary inversely with the proportion of output accounted for by tradable inputs and the nominal rates of the trade restrictions facing them, but directly with the nominal tariff or nontariff rate on final imports.

With respect to duties, the relevant factors are embodiedin the concept of effective rates of protection developed by Johnson, Corden, and others.[5] If input coefficients are constant in the relevant range, the effective rate of duty (z_j) for any commodity (j) can be expressed in the framework of an input-output system.[6] Let (t) denote the nominal rate of tariffs, (a) the material input coefficient, and (v) the proportion of value-added to output, all measured at world market prices. For commodity (j) we have, then,

$$z_j = (t_j - \Sigma\, a_{ij}t_i)/v_j \qquad (2.1)$$

Observing the behavior of the formula, one notes that effective tariff rates (z_j) are highest when input duties (t_i) are zero. Elimination of negligible input duties, however, will cause no change in domestic supply. In this case, the estimating procedure based on nominal rates, which we employin Chapter 4, retains its ability to make valid predictions about changes in the value of imports. On the other hand, when t_i is positive, z_j becomes smaller, and eliminating t_i can result in a substantial increase in domestic supply. Johnson has demonstrated[7] that if input duties (t_i) are lowered along with t_j, the production effect of lowering t_j will be more or less offset by the increased supply following the reduction of t_i. As a result, the increase in imports of (j), when both t_j and t_i go to zero, will be smaller. Indeed, if t_i is large enough to make $z_j = 0$, the value of imports will be unchanged by an across-the-board removal of t_i and t_j. When z_j is negative, removing t_j and t_i will cause the value of imports to decrease. Since it seems most reasonable that t_i is a positive number, estimates in Chapter 4 likely will be upper limits of the actual value of imports excluded by any given trade restriction. Therefore, after calculating the restrictiveness of nominal tariffs and nontariff barriers, we will consider their sensitivity to changes of input restrictions.

Evaluating Effective Tariffs

We can also use nominal and effective tariff rates generated by Basevi and Balassa to evaluate various claims by the AISI, and we employ Basevi's methods to evaluate the effects of the Dillon Round on United States tariff rates.

An International Comparison. According to the AISI, an undocumented study found that "American tariffs on steel are about half as great as European tariffs."[8] However, neither the nominal post-Kennedy Round rates of Table 2-1 or of the Senate Finance Committee substantiate that claim, nor do Balassa's post-Dillon Round rates.

Balassa presents 1962 nominal tariff rates on 36 categories of manufactures including steel for the United States, Japan, Sweden, United Kingdom, and the European Economic Community (EEC).[9] He obtained the rates from *Comparative Tariffs and Trade*,[10] adjusted them for tariff concessions made in the Dillon Round, and averaged the rates with weights determined by the total imports of the five areas.

Because comparable input-output tables for these countries were unavailable, Balassa used "standardized" input-output coefficients derived from Belgian and Dutch tables. He chose these countries' coefficients because they had nil or low duties on most commodities in 1959. Hence, he avoided the distortion in input-output relationships which might arise from existence of duties. As Balassa points out, the use of common input-output coefficients can be justified if the countries have identical production functions with unitary substitution elasticities in all industries, or if national disparities in efficiency are neutral in the sense that production functions vary only by a multiplicative constant. According to these assumptions, divergence in the relative prices of inputs would not affect the coefficients.

While these assumptions do not conform strictly to reality, we can tolerate nonneutral differences in production functions since companies in these industrial countries probably operate under the same "technological horizon." However, in the case of steel some evidence suggests that the monopolistic structure of the United States industry may have hindered its technological development relative to Europe's. To the extent that this is true, using standardized coefficients may distort input-output relationships more than the existence of duties does.

In general, Balassa found effective tariffs to be higher than nominal duties, especially in the cases of textiles, leather, some chemical materials, nonferrous metals, and steel ingots. For the steel industry in particular, we show his results in Table 2-3, together with a weighted average of each area's steel rates. Both the EEC and Sweden impose lower average nominal duties than the United States. Those of the Inner Six, however, undercut the United States average by only 1 percent. Turning to the more relevant effective tariff rates, we find the United States in possession of the highest rates on ingots and the lowest on rolling mill and other steel products. For the steel industry as a whole, the average effective rates lie above those of Sweden, the EEC, and even Japan, on the one hand, and below those of the United Kingdom on the other. Thus the AISI assertion seems to lack support with regard to nominal and effective rates of protection.

An Interindustry Comparison The AISI also asserts that American tariffs in steel run from "one-half to one-third of those for other industrial inputs."[11] To verify this declaration for both nominal and effective tariffs, we can turn to work done by both Balassa and Basevi.

Table 2-4 shows the position of steel manufactures in Balassa's ranking of 36 categories of manufactures according to height of duty. He constructed this industrial ranking for each of the five industrial areas he considered. In addition, we can observe the average rank of the steel industry as a whole for each country and country group. Lastly, Table 2-4 gives the simple average duty on steel and on the entire 36 classes of manufactures.

From Balassa's rankings we can see that effective U.S. steel tariffs range from the highest of the 36 categories on ingots, to next to the lowest on rolling mill products. Nominal rates in the United States, however, tend to cluster at the lower end of the scale, and as the AISI claims, the steel average equals about half of the all-industry average. However, as the average ranks of nominal steel duties indicate, every country places the industry's products at strikingly similar (low) positions in their statutory tariff structures. With regard to effective tariffs, United States steel import duties rank at an average of 24, also a low position on the scale of tariff heights. The Common Market duties occupy the same relative position as those of the United States, while the British and Japanese steel duties climb higher, on the average, in their tariff structures, and the Swedish lower at 29. Within the United States, Table 2-4 also shows that the average effective tariff rate for steel lies slightly above the 36 industry average. Thus while the AISI correctly characterized nominal rates à la Balassa, the same cannot be said for effective rates.

Balassa's calculations, however, do not receive complete confirmation from Basevi's work, although we may expect that the differences arise from Basevi's use of 1954 and 1958 input-output coefficients, tariffs, and f.o.b. import values supplied by the United States Bureau of the Census.[12]

Table 2-5 gives Basevi's results for each of five (three in 1954) census

Table 2-3
1962 Nominal[a] and Effective[b] Tariff Rates[c] on Steel

	U.S.		*U.K.*		*E.E.C.*		*Sweden*		*Japan*	
	t_j	z_j	t_j	z_j	t_j	z_j	t_j	z_j	t_j	z_j
Pig iron and ferromanganese	1.8	9.3	3.3	17.9	4.0	−13.8	0.0	− 0.7	10.0	54.3
Ingots and other primary steel forms	10.6	106.7	11.1	98.9	6.4	28.9	3.8	40.0	13.0	58.9
Rolling mill products	7.1	− 2.2	9.5	7.4	7.2	10.5	5.2	13.2	15.4	29.5
Other steel products	5.1	0.5	17.0	46.8	9.9	20.9	5.4	9.5	13.4	14.1
All steel products[d]	7.7	37.7	12.0	49.3	7.6	19.5	3.6	15.9	14.0	35.5

[a] t_j
[b] z_j
[c]Percentage of c.i.f. value.
[d]Average of other four classes weighted by the share of each commodity group in the total exports of the major steel-trading countries.

Source: Balassa, *Trade Liberalization among Industrial Countries* (New York: Council on Foreign Relations, 1967), pp. 180, 181.

Table 2-4
1962 Ranks of Steel Tariffs[a] among Those of Thirty-six Industries

	U.S.		U.K.		E.E.C.		Sweden		Japan	
	t_j	z_j	t_j	z_j	t_j	z_j	t_j	z_j	t_j	t_j
Pig iron and ferromanganese	35	27	34	28	32	36	36	36	33	6
Ingots and other primary steel forms	22	1	28	1	31	9	28	2	29	5
Rolling mill products	26	35	31	34	30	30	26	23	22	18
Other steel products	31	34	18	3	28	19	25	28	27	31
Average rank of all steel products	29	24	28	17	30	24	29	22	28	15
Average tariff on all steel products[b]	6.1	26.1	10.2	42.7	6.9	11.6	3.6	15.9	13.0	39.2
Average tariff for 36 industries[b]	12.5	22.3	16.2	29.5	12.5	20.2	7.8	16.7	17.8	3.0

Note: t_j = nominal tariff rate; z_j = effective tariff rate.

[a]Percentage of c.i.f. value.

[b]Unweighted average.

Source: Balassa, *Trade Liberalization Among Industrial Countries* (New York: Council on Foreign Relations, 1967), pp. 182, 183.

four-digit steel industries. Also, the simple averages of all steel rates compared with the average tariffs of a large sample of census industries are shown. Referring to column 2 in Table 2-5, we can see that 1954 effective tariffs loom over one-third larger than the statutory rates, while those for 1958-60 are about one-and-a-half times as high as nominal tariff rates.

For the steel industry as a whole, 1954 effective rates were just under one-and-a-third times the nominal tariffs, while in 1958 the five-category industry's effective rate of protection exceeded nominal rates by almost 70 percent. Despite these larger disparities between nominal and effective duties in the steel industry, its average nominal and effective rates in both years amounted to less than the same simple averages for the sample industries. Thus, according to Basevi's figures, nominal steel tariffs do not lie as far below average as the AISI implies, although in contrast to Balassa, Basevi's effective steel tariffs rank slightly below average.

An Intertemporal Comparison Basevi's figures in Table 2-5 show a decrease in nominal steel tariffs between 1954 and 1958 of about 15 percent, but a slight increase in effective rates. The average rates for all industries rose between the two years. This result seems to belie concessions negotiated at Geneva, Switzerland, in the General Agreement on Tariffs and Trade (GATT), dated May 23, 1956.[13] However, Basevi rejects the use of his results to constrast tariffs over time because of differences in sampling procedures.

A glance at the nominal tariff rates in Tables 2-1 and 2-6 does confirm that United States steel tariffs fell as a result of Dillon Round concessions formally completed in July 1962, and Kennedy Round concessions made in 1967. However, since negotiations do not always lower input and output tariffs commensurately, effective protection of the steel industry sometimes falls more than nominal rate decreases would suggest. A comparison of nominal and effective rates between 1958 and 1963 will suffice as an illustration.

Comparable 1958 and 1963 input-output data,[14] nominal tariff rates[15] and weights based on the value of total United States shipments in each four-digit industry provide the raw material for our calculations. We follow Basevi's procedure, with the exception of assigning 1958 (1963) all-industry average tariffs of 0.062 (0.072) to 1958 (1963) material residuals. Output residuals were tariffed at the unweighted average of duties on all other products in each industry classification. Where imports were absent, the commodity was tariffed at rates derived from *Schedule A: Statistical Classification of Commodities Imported Into the United States* effective June 30, 1958, and July 1, 1963.[16]

Table 2-5
Selected U.S. Tariff Rates and Rates of Effective Protection (Percentage)

	1958			*1954*		
	(1)	(2)	(3)	(1)	(2)	(3)
Blast furnaces and steel mills	6.1	8.5	10.5	NA	NA	NA
Gray-iron foundries	8.0	9.7	11.3	9.0	10.2	11.2
Malleable iron foundries	18.2	29.4	32.2	18.2	26.3	27.7
Steel foundries	9.0	11.4	13.3	11.5	14.8	16.3
Iron and steel forgings	13.1	27.3	29.1	NA	NA	NA
All steel (unweighted average)	10.9	17.3	18.3	12.9	17.1	18.4
All industries (unweighted av.)	14.1	20.8	24.5	13.9	18.9	21.9

Notes: (1) Nominal advalorem tariff.

(2) Effective tariff (t_r = 7.0; 5.1).

(3) Effective tariff (t_r = 1.0)

t_r Average rate applied to miscellaneous materials.

Source: Based on figures from G. Basevi, "The United States Tariff Structure: Estimates of Effective Rates of Protection of United States Industries and Industrial Labor," *Review of Economics and Statistics* (May 1966), pp. 153, 155, 156.

Table 2-6 presents nominal and effective tariffs calculated with Basevi's formula. With the exception of gray iron foundry products, all steel output had lower nominal rates of duty in 1963 than in 1958. Material duties, however, went up. This result may be a reflection of the rise in average duties collected on all imports during the period, since this figure applied to the large miscellaneous materials categories in the input-output tables. In addition, increases in many tariff rates on metals used as alloys in steel production occurred. Nominal output duties, averaged for the entire steel industry, fell 26 percent while an industry-average of statutory input duties rose 23 percent.

As expected, both of the above trends combined to depress the steel industry's average effective duty by 54 percent. Once again, gray iron foundry products, alone, escaped a decrease, while malleable iron foundries sustained the most dramatic decrease—71 percent. The reasons underlying such different results between the two foundry categories appear to be the greater proportion of nonferrous metals and miscellaneous materials used in the latter, as well as a 64 percent decline in the nominal tariff on malleable iron castings.

Nontariff Barriers

As Chapter 1 notes, we cannot record even the nominal rates of nontariff

Table 2-6
United States Nominal and Effective Tariff Rates on Steel (Percentage)

	1963			*1958*		
	(1)	(2)	(3)	(1)	(2)	(3)
Blast furnaces and steel mills	7.0	3.6	11.3	8.8	3.7	16.3
Gray-iron foundries	9.1	2.6	14.1	7.1	3.2	10.3
Malleable iron foundries	8.0	4.9	9.7	19.9	4.2	33.5
Steel foundries	6.0	5.8	6.0	10.8	5.0	14.1
Iron and steel forgings	6.3	9.9	2.3	11.7	5.5	19.7
All steel (unweighted average)	7.2	5.3	8.6	11.6	4.3	18.7

Notes: (1) Nominal advalorem tariff on output.

(2) Nominal advalorem tariff on inputs.

(3) Effective tariff (t_r = 7.2; 6.2).

t_r Same as in Table 2-5.

Source: See text.

restrictions in the relatively small space accorded to tariffs. Therefore we leave detailed discussion of nontariff barriers to Part II. Chapters 5 to 8 in Part II indicate that the potential restrictiveness of each of these barriers varies markedly with the product and country under consideration. Nevertheless, we can make some useful generalizations about the number and type of nontariff restrictions in each of the ten major steel-trading countries.

The Distribution of Nontariff Barriers

Figure 2-1 shows that nontariff restrictions generally imposed on all imports concentrate in the areas of customs administration and subsidies. Very few countries try to exclude imports across the board through the use of any foreign trade policies except export assistance. Likewise, very few employ internal policies (outside of domestic subsidies) or permit restrictive business practices to affect all imports.

West Germany and Canada apply the least number of general restraints (7). France and the United Kingdom set up the highest number (13) of barriers against all goods. Only Japan requires licensing for all imports, while only Luxembourg levies an import surcharge on all goods. Not one of the major steel-trading countries imposes quotas, "voluntary" export limitations, biased government procurement, or antidumping measures on all imports. Similarly, no one attempts generally to deter imports with advertising or marking restrictions.

Figure 2-2 shows that countries favor foreign trade policies (with the exception of export subsidies), marking, technical, health, and safety regulations, as well as restrictive business practices for impeding imports on a selective basis. They seldom impose discriminatory customs practices, domestic subsidies (with the exception of low-cost loans and overpriced government purchases), and national security restrictions on selected imports only. Japan erects the smallest number (11) of selectively applied nontariff barriers, while France imposes the most (15). Germany and the United States account for 14, while Canada has 11, and the rest have 13. France is the only major steel trader to restrict some imports with advertising restrictions, while the United Kingdom is the only one to give a rebate on internal taxes as a subsidy to selected goods.

Figure 2-3 shows the distribution of each major steel trader's nontariff barriers to steel imports. Steel is subject to 14 of the 20 selective restrictions, as well as to all barriers which are imposed across the board, and the three restrictions which apply both ways. Thus one or another of the major steel-trading countries restricts steel imports with 34 different nontariff barriers. The only barriers that steel avoids to any extent are those erected by internal policies. Steel faces no advertising restrictions, direct tax or other internal tax impediments, credit controls, patent regulations, health and safety rules, or national security restrictions. France imposes the most restrictions on steel (21) and Canada the least (10). Germany, Luxembourg, and the Netherlands each account for 15 steel restrictions, while Belgium and Italy have 16. Japan applies 18 barriers, while the United States and the United Kingdom impose 17 each.

Quantifiable Nontariff Barriers to Steel

Chapter 1 noted that this study intends to estimate the restrictive effects of the nontariff barriers listed in Figure 2-3. However, we can make precise estimates only if we can determine the height of the barrier. In other words, the import-domestic price differential created by each nontariff restriction must be obtainable. As we shall see in the next section of this chapter, we can express the differential at an ad valorem tariff-like rate, and we can use these rates as well as elasticity estimates (from Chapter 3) to predict restrictiveness.

Unfortunately, a glance at Figure 2-3 confirms the earlier observation in Chapter 1 that we cannot express many important steel restrictions in numerical terms. In particular, most licensing and exchange controls, prejudice in government procurement, antidumping laws, administrative practices, noneconomic internal policies, and some restrictive business practices are not amenable to transformation into a tariff-equivalent rate.

		Belgium	Canada	France	Germany	Italy	Japan	Luxembourg	Netherlands	United Kingdom	United States
Foreign Trade Policies	Import licenses						X				
	Exchange controls			X			X		X		
	Quotas										
	Export limitations										
	Biased procurement										
	Transport rebates	X		X		X		X	X		
	Wage tax rebates			X			X				
	Marketing assistance	X	X	X	X	X	X	X	X	X	X
	Tied foreign aid			X							X
	Indirect tax rebates	X				X		X	X	X	
	Import surcharges							X			
	Antidumping laws										
Administrative Practices	Advertising rules										
	Marking regulations										
	Document requirements	X	X	X	X	X	X	X	X	X	X
	Classification rules	X	X	X	X	X	X	X	X	X	X
	Valuation procedures	X	X	X	X	X	X	X	X	X	X
	Customs fees & deposits			X		X	X			X	
	Other customs rules	X	X	X	X	X	X	X	X	X	X
	Penalties	X	X	X	X	X	X	X	X	X	X
Internal Policy & Regulation	Direct payments	X								X	
	Depreciation	X	X	X	X	X	X	X	X	X	X
	Investment incentives	X		X		X	X	X	X	X	X
	Low cost loans										
	Overpriced purchases										
	Internal tax rebates										
	Direct taxes										
	Indirect taxes	X				X			X		
	Other internal taxes										
	Price controls										
	Credit controls										
	Advertising rules										
	Patents & trademarks										
	Health & safety rules										
	Technical specifications										
	National security rules										
Private Practices	Cartels										
	Exclusive supply agreements										
	Freight rate discrimination										

Source: See text.

Figure 2-1 Nontariff Barriers Generally Imposed on All Imports

		Belgium	*Canada*	*France*	*Germany*	*Italy*	*Japan*	*Luxembourg*	*Netherlands*	*United Kingdom*	*United States*
Foreign Trade Policies	Import licenses	X	X	X	X	X		X	X	X	X
	Exchange controls	X				X		X	X		
	Quotas	X	X	X	X	X	X	X	X	X	X
	Export limitations		X	X	X					X	X
	Biased procurement	X	X	X	X	X	X	X	X	X	X
	Transport rebates										
	Wage tax rebates										
	Marketing assistance										
	Tied foreign aid										
	Indirect tax rebates										
	Import surcharges										
	Antidumping laws	X	X	X	X	X	X	X	X	X	X
Administrative Practices	Advertising rules			X							
	Marking regulations	X	X		X	X	X	X	X	X	X
	Document requirements										
	Classification rules										
	Valuation procedure										
	Customs fees & deposits										
	Other customs rules										
	Penalties										
Internal Policy & Regulation	Direct payments										
	Depreciation										
	Investment incentives										
	Low cost loans			X	X		X				
	Overpriced purchases	X			X	X		X	X		
	Internal tax rebates									X	
	Direct taxes										
	Indirect taxes		X								
	Other internal taxes			X							X
	Price controls	X		X	X	X	X	X	X	X	
	Credit controls									X	X
	Advertising rules			X						X	X
	Patents & trademarks										X
	Health & safety rules	X	X	X	X	X	X	X	X		
	Technical specifications	X	X	X	X	X	X	X	X		
	National security rules										X
Private Practices	Cartels	X	X	X	X	X	X	X	X	X	X
	Exclusive supply agreements	X	X	X	X	X	X	X	X	X	X
	Freight rate discrimination	X	X	X	X	X	X	X	X	X	X

Figure 2-2 Nontariff Barriers Selectively Imposed on Some Imports

		Belgium	Canada	France	Germany	Italy	Japan	Luxembourg	Netherlands	United Kingdom	United States
Foreign Trade Policies	Import licenses						X				
	Exchange controls			X			X				
	Quotas			X							
	Export limitations										X
	Biased procurement	X	X	X	X	X	X	X	X	X	X
	Transport rebates	X		X	X	X		X	X		
	Wage tax rebates			X							
	Marketing assistance	X	X	X	X	X	X	X	X	X	X
	Tied foreign aid			X							X
	Indirect tax rebates	X			X	X		X	X	X	
	Import surcharges							X			
	Antidumping laws		X		X						X
Administrative Practices	Advertising rules										
	Marking regulations										X
	Document requirements	X	X	X	X	X	X	X	X	X	X
	Classification rules	X	X	X	X	X	X	X	X	X	X
	Valuation procedures	X	X	X	X	X	X	X	X	X	X
	Customs fees & deposits			X		X	X			X	
	Other customs rules	X	X	X	X	X	X	X	X	X	X
	Penalties	X	X	X	X	X	X	X	X	X	X
Internal Policy & Regulation	Direct payments	X								X	
	Depreciation	X	X	X	X	X	X	X	X	X	X
	Investment incentives	X		X		X	X	X	X	X	X
	Low cost loans			X	X		X				
	Overpriced purchases			X							
	Internal tax rebates									X	
	Direct taxes										
	Indirect taxes	X	X			X			X		
	Other internal taxes										
	Price contros controls	X		X	X	X	X	X	X	X	X
	Credit controls										
	Advertising rules										
	Patents & trademarks										
	Health & safety rules										
	Technical specifications			X							
	National security rules										
Private Practices	Cartels	X		X	X	X	X	X	X	X	X
	Exclusive supply agreements	X		X	X	X	X	X	X	X	X
	Freight rate discrimination						X				

Source: See text.

Figure 2-3 Nontariff Barriers Facing Steel Imports

The nontariff barriers which we can quantify appear in Table 2-7. As illustrated there, we find six quantifiable nontariff restraints in both France and Italy; five in Belgium, Japan, Luxembourg, the Netherlands, the United Kingdom, and the United States; four in Canada; and two in Germany. Of course, for any one country the number of barriers we can quantify may bear no relation to its total number of nontariff restrictions, nor to their total impact on steel imports.

Effective Rates of Nontariff Barriers

Another qualification facing estimates based on nominal nontariff barriers relates to their failure to account for the effective rate of protection. Johnson demonstrated that the effective rate of tariffs will be lower in the presence of nontariff restrictions applied across the board.[17] For instance, excise taxes levied at the same rate on imports of both output and inputs will reduce the effective-protection rate accorded by tariffs. Johnson does not mention, however, that many nominal nontariff barriers may have higher effective rates themselves, because the restrictions on inputs may be lower.

Whether or not this is the case in the steel industry cannot be determined with any precision. Countries usually subject steel inputs to domestic subsidies, transport rebates, low or nonexistent excise taxes, and exempt them from customs regulations, advertising, patent, trademark, and health or safety rules. They do appear to be restrained by the same price preferences, import surcharges, antidumping laws, customs fees or deposits, penalties, technical specifications, and restrictive business practices imposed on steel output. On the other hand, countries more often subject raw materials to quantitative restrictions, price controls, national security regulations, and other import barriers. Therefore, in the absence of a comprehensive investigation of every barrier erected against steel inputs, we find it difficult to determine what difference exists between the nominal and effective rate of protection afforded by nontariff restrictions on steel.

It is nevertheless quite obvious that nontariff barriers on steel inputs have a positive rate. Thus estimates based on nominal rates will overstate the effects of across-the-board reductions in nontariff barriers. We will assess the degree of this overstatement in Chapter 4.

The Restrictive Effect of Trade Barriers

The way in which tariffs and quantified nontariff barriers restrict imports

Table 2-7
Quantified Items Creating Nontariff Barriers to Steel[a]

Nontariff Barrier Items	*Major Steel-trading Countries*									
	Belgium	Canada	France	Germany	Italy	Japan	Luxem-bourg	Nether-lands	United Kingdom	United States
Quotas			59,000[b]							
"Voluntary" export limits										14,000,000[c]
Domestic biased purchasing	10.00%	10.00%[f]					10.00%	10.00%		6.00%[d]
Import surcharges							3.00%			10.00%
Export subsidies:										
Direct payment			6.00%							
Credit financing			2.70%		4.70%	1.20%				
Penalties						$28[g]				
Tax rebates	7.00%				22.00%		1.00%	5.57%	3.00%	
Prior deposit requirements						.09%			1.90%	
Document costs	3.37%	3.37%	3.37%	3.37%	3.37%	3.37%	3.37%	3.37%	3.37%	3.37%
Customs fees			.25%		.60%					
Domestic subsidies:										
Direct Payments									8.00%	
Internal tax rebate									.60%	
Input subsidies	1.30%	3.40%	1.00%	1.20%	0.80%		1.30%	1.30%		
Indirect taxes	6.58%	12.00%[e]			4.80%			5.75%		
Freight rate discrimination						.49%				3.26%

Notes:

[a]Amount of quota or percent of c.i.f. import price.

[b]Metric tons per month.

[c]Short tons per year.

[d]6% for federal purchases other than for Department of Defense (28%), and AID which has limited its foreign purchases of steel to 7% per year.

[e]Applies to 80% of Canadian steel imports.

[f]Applies to Provincial as well as federal purchases.

[g]Per ton of production.

Source: See text.

often determines their height. This section analyzes the restrictiveness of various types of barriers in order to obtain a measure of their ad valorem rates.

Tariffs and Quotas

Any import restriction causes a decrease in the effective supply of goods to importers or alternatively a decrease in the effective import demand facing foreign exporters. According to the former interpretation, a tariff with specific rate T would cause import prices to rise by dP', a fraction (s) of T while prices received by suppliers will fall by $dP = (1-s)T$ as long as the absolute values of import demand and supply elasticities (η, ϵ) lie between zero and infinity. Therefore, the percentage increase in import prices including duty will equal

$$\begin{aligned} dP'/P &= sT/P \\ (M/dM)(dP'/P) &= st(M/dM)(P/dP)(dP/P) \text{ where } t = T/P. \\ 1/\eta &= stP[\epsilon(1 - s)T]^{-1} \\ \epsilon(1 - s)t &= st\eta \\ \epsilon t &= st\eta + st\epsilon \\ dP'/P &= t\epsilon/(\eta + \epsilon) \end{aligned}$$

Likewise, the duty-free price will fall by the following percentage:

$$\begin{aligned} dP/P &= (1 - s)t \\ &= t\eta/(\eta + \epsilon) \end{aligned} \qquad (2.3)$$

Using d to denote a small change, we can write the decrease in the quantity of imports (M) after imposition of the tariff as

$$\begin{aligned} dM &= M(dM/M)\,(P/dP')\,(dP'/P) \\ &= Mt\eta\epsilon/(\eta + \epsilon) \end{aligned} \qquad (2.4)$$

The decrease in the duty-free value of imports ($V = MP$) is

$$\begin{aligned} dV &= PdM + MdP - dPdM \\ &= PdM + MPdP/P - PdMdP/P. \end{aligned} \qquad (2.5)$$

Substituting (2.3) and (2.4) into (2.5), we obtain

$$dV = Vt\eta[\epsilon + 1 - t\eta\epsilon(\eta + \epsilon)^{-1}] (\eta + \epsilon)^{-1}. \qquad (2.6)$$

Whenever $\epsilon = \infty$, $dP'/P = t$, $dP = 0$, and the change in the value of imports depends on η:

$$dV = PdM \qquad (2.7)$$

$$= Vt\eta$$

Whenever $\eta = \infty$, $dP' = 0$, $dP/P = t$, and the duty-free value of imports will fall by

$$dV = Vt(\epsilon + 1 - t\epsilon). \qquad (2.8)$$

Since we usually cannot observe pretariff values of V, P, or M, however, equations (2.6, 7, 8) possess limited applicability. To arrive at practical estimates of restrictiveness, we observe the duty-free V, M, and P subjected to tariff rate t, and calculate the increase (dV) that would follow elimination of the trade restrictions. In this case, doing away with t will cause the import price to decline by the following percentage:

$$\mathrm{d}P'/(P + T) = st/(1 + t)$$

$$= t\epsilon/(\epsilon + t\epsilon + \eta). \qquad (2.9)$$

Similarly, the duty-free price will rise by

$$dP/P = (1 - s)t$$

$$= t\eta/(\epsilon + t\epsilon + \eta). \qquad (2.10)$$

As a result, the percentage increase in import volume which will follow elimination of t becomes

$$dM/M = t\eta\epsilon/(\epsilon + t\epsilon + \eta). \qquad (2.11)$$

Generally, the increase in the duty-free value of imports after removal of t equals

$$dV = MdP + PdM + dPdM$$

$$= V[(1 + dP/P)(1 + dM/M) - 1]. \qquad (2.12)$$

Substituting (2.10) and (2.11) into (2.12), one obtains

$$dV = Vt\eta[1 + \epsilon + t\eta\epsilon(\epsilon + t\epsilon + \eta)^{-1}](\epsilon + t\epsilon + \eta)^{-1}. \qquad (2.13)$$

Whenever $\epsilon = \infty$,

$$dV = Vt\eta/(1 + t). \qquad (2.14)$$

Whenever $\eta = \infty$,

$$dV = Vt(1 + \epsilon + \epsilon t). \qquad (2.15)$$

Clearly, the imposition of any quota reducing import volume by dM in (2.4) would possess the same restrictiveness as an ad valorem tariff with rate t. However, demand growth may occur during the period covered by the quota. Thus we obtain a more accurate estimate of restrictiveness by subtracting the actual values imported under the quota (V) from the potential value of unrestricted imports (V'). We can derive the latter figure from income elasticities of demand, observed prequota import values (V'') and volume (M''), and price elasticities for import demand and supply.

Growth in import demand by dM/M'' would create a price gap at M'' (expressed as a percentage of the prequota price) of $(dM/M''\eta)(1 - dM/M''\eta)^{-1}$. The potential import value (V') will equal V'' plus dV from equation (2.13) with the price gap and V'' substituted for t and V. Whenever $\epsilon = \infty$, V' equals V'' $(1+ dM/M'')$. Whenever $\eta = \infty$, we need only compare V'' and V to determine the restrictiveness of the quota.

Direct Nontariff Barriers

Other nontariff restraints produce the same effect as tariffs or quotas, although the changeable and covert nature of such restrictions often creates enough uncertainty to accentuate their restrictiveness. The costs of obtaining import licenses and exchange permits, price preferences granted in government procurement, countervailing duties, surcharges, statistical or handling fees, custom's deposits, fines, costs of satisfying certain customs and/or internal regulations, excess border taxes, and freight rate differentials—all have precisely the same effect as a duty of equal amount. Exchange and import licensing systems state trading, nonprice government procurement preferences, and national security restrictions create an obviously quota-like effect on imports, though they have not been quantified.

Subsidies

Tariff-like restrictions of imports also result from subsidies, if import demand can be looked upon as the difference between domestic demand and supply at prices below domestic equilibrium. In that case, any specific or ad valorem subsidy which increases domestic supply also decreases import demand. However, the rate of subsidization need not equal the ad valorem rate of the nontariff barrier created by the subsidy because of incidence considerations.

Domestic Subsidies. A domestic subsidy will lower import demand only to the extent it is passed on to buyers in the form of a price decrease. According to orthodox economic theory, a lump-sum domestic subsidy would have no immediate effect on domestic supply, but a domestic subsidy (with ad valorem rate b) would increase domestic supply and decrease the domestic equilibrium price in the absence of trade by a fraction of the subsidy. In percentage terms, we can derive the amount of the price decrease in a manner analogous to the incidence analysis of equation (2.2). Substituting b for t, and domestic supply and demand elasticities (ϵ_D and η_D) for the corresponding import elasticities, we obtain the following expression for the percentage price decrease: $b\epsilon_D/(\eta_D+\epsilon_D)$. We can substitute this tariff equivalent nontariff barrier rate for t in equations (2.13), (2.14) and (2.15) to determine the restrictive effect of domestic subsidies.

As specified by the expression, the rate of domestic subsidization equals the rate of the nontariff barrier it creates when the domestic supply elasticity equals infinity or the domestic demand elasticity equals zero. On the other hand, domestic subsidization does not create any nontariff barrier when domestic demand is infinitely elastic or when the supply elasticity is zero.

Export Subsidies. We cannot use traditional international trade theory to show how an export subsidy erects an import barrier, because its framework cannot accommodate a country which imports and exports the same good. Nevertheless, in the real world many countries trade both ways with the same commodity. All of the major steel-trading countries, for instance, import and export a substantial amount of iron and steel mill products. Part of this two-way trade results from product differentiation, imperfect knowledge, international versus intranational transport costs, institutional arrangements such as long-term contracts or other ties between buyer and seller, and intranational factor immobility; however, perfectly competitive long-run equilibrium in the domestic market at a given free-world trade price[18] may also result in two-way trade. In the absence of transportation costs, domestic consumers will be indifferent between imports and domestic goods, while domestic producers will be indifferent between selling domestically or exporting. In this case an export subsidy will increase the country's production and exports, but need not change the volume of imports.

Suppose, on the other hand, that the subsidizing country faces a less than infinitely elastic demand for its exports. Then the export subsidy will lower export prices to the extent that elasticities of demand (η_x) and supply (ϵ_x) for the subsidized export permit it to be passed on to foreign buyers. In turn, the possibility of reimportation of the subsidized export and the prevalence of antidumping laws (discussed in Chapter 5) ensure that the domestic price will also fall to equality with the new export price.[19]

When the domestic equilibrium price falls, the import demand function facing foreign sellers also decreases. As a result, the export subsidy creates a nontariff barrier, equal to that part of the export subsidy passed forward to buyers. According to the method previously employed, the nontariff barrier as a percentage of c.i.f. import price will equal $\epsilon_x/(\eta_x+\epsilon_x)$ times the rate of export subsidization. With values for ϵ_x and η_x from the following chapter, we can convert export subsidies to their ad valorem tariff-equivalent rates in Chapter 4. The latter chapter, of course, will also present estimates of the value of steel restricted by all import barriers.

Notes

1. AISI, *The Steel Import Problem*, p. 32.

2. Kelly, "Nontariff Barriers," p. 286.

3. Joint Economic Committee, 89th Cong., Hearings before the Subcommittee on Foreign Economic Policy: *The Future of United States Foreign Trade Policy* (Washington, D. C.: United States Government Printing Office, July 1967), p. 331.

4. Senate Finance Committee, *Steel Imports*, p. 48.

5. See for instance, H. Johnson, "The Theory of Tariff Structure, with Special Reference to World Trade and Development," *Trade and Development*, Etudes et Travaux de l'Institut Universitaire de Hautes Etudes Internationales, no. 4 (Geneva 1965), p. 10.

6. The assumptions underlying the use of an input-output framework in the model have been summarized by J. Leith, "Substitution and Supply Elasticities in Calculating the Effective Protective Rate," *Quarterly Journal of Economics* (November 1968), p. 589:

(1) The rate of duty expresses the rate of divergence between protected and free-trade prices of a tradable.

(2) The production function describing the relationship between inputs and outputs is linearly homogeneous.

(3) The elasticities of substitution between inputs is zero.

(4) The elasticity of foreign supply of exports facing domestic buyers is infinite.

(5) The elasticity of supply of domestic nontradable inputs is infinite.

(6) The elasticity of supply of other factors of production to the domestic industry is less than infinite.

(7) Production and trade are present before and after the introduction of protection.

In the same contribution Leith proceeded to determine the sensitivity of the effective rate of protection to relaxation of some of these restrictive assumptions. In particular, a nonzero substitution elasticity, or a less than infinite supply of inputs appears to reduce the effective rates.

7. H. Johnson, "The Theory of Effective Protection and Preferences," *Economica* (May 1969), p. 124.

8. C. Cotton, "GATT for Equality in World Trade," *Steelways* (March 1964), pp. 2-7.

9. B. Balassa, *Trade Liberalization among Industrial Countries*, (New York: Council on Foreign Relations, 1967), p. 49.

10. F. Topping, *Comparative Tariffs and Trade*, New York: Committee for Economic Development, 1964.

11. Cotton, "GATT."

12. G. Basevi, "The United States Tariff Structure: Estimates of Effective Rates of Protection of United States Industries and Industrial Labor," *Review of Economics and Statistics* (May 1966), p. 148.

13. United States Department of Commerce, *Schedule A: Statistical Classification of Commodities Imported into the United States*, Washington, D. C.: United States Government Printing Office, 1958.

14. United States Department of Commerce, Bureau of the Census, *Census of Manufactures*, Washington, D. C.: United States Government Printing Office, 1958, 1963.

15. United States Department of Commerce, Bureau of the Census, *U. S. Commodity Exports and Imports as Related to Output*, Washington, D. C.: United States Government Printing Office, 1958, 1963.

16. Department of Commerce, *Schedule A.*, 1958, 1963.

17. Johnson, "Theory of Effective Protection."

18. The given world price implies that import supply and export demand are infinitely elastic for the country under consideration.

19. Towle, *International Trade and Commercial Policy*, p. 471.

3 Demand and Supply Elasticities

In order to evaluate the effects of tariffs and nontariff barriers on trade flows among the major steel-trading countries, we need to know their import demand and supply elasticities. As we also learned in Chapter 2, converting export subsidization rates to nontariff barrier rates necessitates the use of export demand and supply elasticities. Domestic demand and supply elasticities are also required to compute the height of the nontariff barriers raised by domestic subsidies. This chapter presents estimates of the appropriate elasticities for each of the major steel traders.

United States Import Demand Elasticity

The problem confronted in estimating the elasticity of demand for steel imports lies in choosing an appropriate model for the import demand function. Rippe formulated a model to explain American steel import shares in a recent *Review of Economics and Statistics* article.[1] However, it possesses some severe a priori limitations which may restrict predictive accuracy and its applicability in this study. It ignores the impact of domestic income changes; it aggregates pig iron (a raw material) with finished steel products; it uses a pricc index that does not cover several important imports (namely pig iron, iron and steel scrap, ferro alloys, wire, nails, cold rolled strips, bars, and other shapes, pipes and tubes) nor any steel from Japan. It also relies on a lagged import variable to explain much of the variance in imports.

An alternative formulation employs a dependent variable that excludes pig iron, an import price index which covers all American steel imports (except pig iron) regardless of type or source, a domestic income variable, and variables to reflect the impact of domestic labor disputes and capacity utilization. Stating the variables in logarithmic form to facilitate forecasting with elasticities we obtain the following equation:

$$M = \alpha + \beta_1 P + \beta_2 Y + \beta_3 C + \beta_4 L_1 + \beta_5 L_2 \qquad (3.1)$$

M = log of import volume index;

P = log of relative price index;

Y = log of real GNP index;

C = log of capacity utilization;

$$L_1 = \text{binary variable} \begin{cases} = 1 \text{ in periods of hedge buying or strikes} \\ = 0 \text{ in other quarters;} \end{cases}$$

L_2 = percentage of man-days lost due to maritime industry strikes.

Ball and Marwah applied the basic procedure used here for five broad commodity groups over the years 1948 to 1958. Kreinin also used it to estimate the import demand elasticity for the same range of goods in the years 1954 to 1964.[2]

Kreinin lagged import volume one quarter behind the two independent income and price variables. This seems to be an appropriate method to account for the time between the consumers' reaction to a change in the import price of a finished manufacture and the resulting level of imports. However, in the market for a capital good such as steel, the consumer and supplier usually keep in closer touch, with fewer intermediaries such as wholesalers and retailers. Here the prices quoted during a particular quarter probably determine the amount of import orders in the same quarter, and an unlagged regression would seem most valid. To determine the effects of lags, we ran the present study's regression with import volume lagged one quarter behind price and income.[3] In the lagged equation, no significant correlation existed between imports and relative prices, the price elasticity took a positive value, and the multiple correlation coefficient was relatively low. Therefore, we employ unlagged estimates for price and income elasticities.

Variables in the Model

Imports

The import volume index comes from data for SITC 67 (iron and steel mill products) published by the United States Department of Commerce.[4] The quarterly figures for SITC 67 are reduced by subtracting SITC 671 (pig iron). See Table A-1 of the appendix.

Prices

Import Unit Values. The relative price index equals the ratio of import to domestic prices. The relevant data appear in Table A-2. Department of Commerce unit values for steel imports[5] provide a measure of import prices that is more complete in its coverage than the continental European export price. Kravis and Lipsey, in a National Bureau of Economic Research (NBER) study, warn that aggregation problems reduce the certainty that changes in unit values represent changes in prices. Nevertheless, they also state that steel products present fewer problems for the construction of unit value indexes than most manufactured goods because the quantity and value data are adequate and because the relative homogeneity of steel products lessens statistical ambiguities arising from quality changes. Deviations of unit values from the Kravis-Lipsey price index are not large. Unit values, for steel, "appear to distort the size, but not the direction, of changes in price competitiveness."[6] For the six periods they considered, the greatest disparities between the unit value and NBER indexes occurred in 1953-57, when the unit value index climbed by six percentage points more. Two other periods exhibited no differences; all deviations were consistently biased upward.

Domestic Prices. Deriving an index of relative prices requires dividing the import price index by the domestic price index for steel. The BLS publishes quarterly domestic prices in composite indexes for all iron and steel, as well as for the following subclassifications in this category: iron ore, pig iron and ferroalloys, iron and steel scrap, semifinished steel, finished steel, and foundry and forge products. The closest approximation of the import definition for iron and steel mill products appears to be a combination of the last three wholesale price index (WPI) subcategories (codes 1013, 1014, and 1015 respectively).[7]

Our elasticity estimates cover the fifteen years between 1954 and 1968. During this period the WPI weights of semifinished, finished, and foundry and forge products changed four times, and the base year of the WPI changed twice.[8] Consequently, the construction of a domestic steel price index required (1) averaging monthly price indexes for each of the three groups over quarterly periods, (2) applying appropriate WPI weight factors for each time period to the three steel product groups, (3) linking the different weighted price indexes for each period so the weighted prices would be continuous over time, (4) adding the weighted indexes to form a composite domestic steel price index, and (5) adjusting the index to one base period: 1957-59 average = 100.

As many economists note, using the WPI as a measure of domestic steel prices supposedly involves the following problems:

1. The WPI measures quoted and not realized prices.
2. Product mix changes may not be reflected by the WPI because the weight for each component is fixed for long periods.
3. Quality changes are not adequately accounted for.
4. Ths WPI misses some price concessions, extra charges, and ignores freight absorption.

The alleged faults are the subject of some dispute, however. The BLS indicates that most reported prices equal the actual selling prices of representative manufacturers collected by mail questionnaires.[9] WPI weights for steel change about once every four years, and major product mix changes in steel do not appear to occur any more frequently. Adequate accounting for quality changes, while difficult, may be easier for steel since its distinguishing features are measurable and can be specified in detail. Moreover, congressional testimony by a BLS commissioner indicates that the WPI will include any shading of steel prices which becomes widespread or is of long duration .[10]

Empirical tests of the WPI differ in their results, but none seem to validate the popular impression that the WPI for steel is grossly inaccurate. Bailey's study cites the difference between *The Iron Age* composite price and actual mill net yield to United States Steel Company during World War I as an example of the magnitude of off-list selling, but his data also show negligible differences ever since that time. Moreover, *The Iron Age* list prices and prices quoted by the BLS are not necessarily equivalent. In referring to BLS price quotations, Bailey admits that "it cannot be shown conclusively that . . . they generally fail to reflect the true frequency of price change."[11]

Flueck compared the WPI for steel with prices bid on government contracts. Of course, the bid prices involve problems relating to the representativeness of government transactions. Flueck also adjusted the bid prices downward with two deductions: one for $1.05 per 100 lbs . . . suggested to him by government steel buyers and specification experts and another of approximately 15 percent for the so-called average delivered transportation cost added by the government. Since the BLS estimates that freight charges average only 7 percent of delivered prices, it is not surprising that Flueck found selected steel prices paid by government consistently 5.0 to 8.4 percent lower than the BLS quotations. Bid prices also appeared to change slightly more frequently than the WPI: twice instead of once for steel sheet and plate in the mid-1950s, and thirteen instead of five times for sheet in the early 1950s. The magnitude of change in bid price exceeded the WPI change by 0.9 percent to 10.8 percent.[12]

Kravis and Lipsey compared the WPI with their index of actual U.S. export prices.[13] For steel the changes in the two indexes appeared to be similar in direction and magnitude:

	1957/1953	1961/1957	1962/1961	1963/1962	1964/1963	1964/1953
NBER	120	100	98	99	101	118
WPI	127	102	100	100	101	130

McAllister's comparison of annual WPI and census data for steel did "not substantiate the hypothesis that the unit values are more flexible on an annual basis than the WPI prices."[14] Finally, the most recent and comprehensive study of WPI reliability by Stigler and Kindahl finds that in this industry transaction prices moved very closely with the BLS index. In their words, "the BLS and NB [NBER] prices of steel products move together so closely that a description of one is a description of the other . . . the general picture was one of close adherence to quoted prices even for very large buyers of steel."[15]

Income

The demand for steel is derived from consumer demand for goods either made of steel or produced by steel machinery, stored in steel containers and buildings reinforced with steel, or transported in vehicles of steel. Thus, GNP, the broadest indicator of income available, seems appropriate as a basis for the second independent variable in our regression equation. We show quarterly totals at annual rates in constant dollars in Table A-3.[16]

Capacity

One of the basic tenets of traditional international trade theory regards import demand as the difference between domestic demand and supply at prices below domestic equilibrium. In Marshallian partial-equilibrium terms, the demand for imports is excess domestic demand.[17] However, rigid, oligopolistic steel prices in the United States may not reflect domestic market conditions. With inflexible prices, but production below capacity, there may be little incentive to buy abroad because domestic steel is easily obtained. Conversely, when domestic demand stimulates full capacity production, then the absence of rationing by the price mechanism may cause excess demand to spill over into the import sector. To test for the influence of this factor, the percentage of capacity utilization in log form is introduced as a third independent variable.

In the period prior to 1961, we find readily available capacity data for the United States steel industry. We can base all estimates on information from the AISI, appearing in Table A-4. The AISI has qualified these estimates by stating that they represent "net steel capacity," a concept which excludes operating time lost because of rebuilding, relining, repairs, and holiday shutdowns. In 1958 the average deduction by producers for lost capacity was 9.1 percent; in 1960 the figure was 8.7 percent.[18] United

States steelmen further decrease this estimate to what may be called practical capacity. In practice, one steel executive has said, "capacity operation gives us so many problems in the way of maintenance and other fields that we have the feeling that we are doing better at, say, 86 to 90 percent of capacity, than at 100 percent. . . ."[19]

Since 1961 when the AISI discontinued its reports, we have had to rely on several different annual estimates of United States steel capacity. These estimates and their sources appear in Table A-5. The Council of Economic Advisors (CEA) made the only continuous estimates supported by at least two other estimators, and for this reason we employ the CEA figures for the regression model. Early CEA estimates approximate those of the Office of Business Economics and are closest to European Coal and Steel Community figures for 1965 and 1970, as well as Bethlehem's prediction for 1970. In contrast, *Wall Street Journal* capacity data appears high, while the United Nations and Office of Business Economics (in recent years) make lower appraisals. Lower still is *The Iron Age* magazine estimate for 1967 of 155 millions tons, but the credibility of this figure is in doubt. Near the end of 1968, *The Iron Age* claimed the industry was running at 40 to 60 percent of capacity, when fourth quarter production at an annual rate equalled 112 million tons.[20]

The capacity estimates from the AISI and CEA enter in equal quarterly rates for each of fifteen years. Dividing them into quarterly production rates for the same period (see Table A-6) yields a ratio, which (in logarithmic form) comprises the third independent variable in the import demand model.

Labor Disputes

Other market conditions not necessarily reflected by price include labor negotiations and strikes. As we noted in Chapter 1, prior empirical tests of the influence of steel strikes produced varying results. Thus, to see if actual or threatened steel strikes are an important determinant of the level of United States steel imports, the regression equation includes a binary variable taking the value of one in buying periods affected by steel strikes or negotiations and zero in all others. Tables A-7 and A-8 in the Appendix summarize the observations.

Ascertaining the quarters affected by labor disputes first involved scanning records of collective bargaining in the steel industry for information on the length of negotiations, dates of strike deadlines, and the duration of strikes. Next, we used steel industry trade publications, primarily *The Iron Age* to obtain market summaries for the period surrounding each contract expiration.[21] A month before 1954 negotiations began, steel producers saw "a few signs here and there of strike hedging," but it had run its course by the time negotiations started. Similarly, some hedge buying was noted a month before 1956 negotiations started. However, a month-long strike also affected steel purchases during the third quarter of 1956. Buyers were anticipating the 116-day strike of 1959 as early as January, and the influence of the strike on their procurement policies continued into the first quarter of 1960. Despite secret 1962 labor negotiations which were concluded by the end of the first quarter, large orders for second-quarter delivery were already extended as a hedge against strikes. Successful negotiations in 1963 also failed to prevent hedge-buying before agreement was reached in late June. Stockpiling by steel users to insure against a May 1, 1965 strike deadline caused domestic mills to apply quotas on many steel products in January of that year. Lengthy negotiations and postponements of the strike deadlines resulted in continuous hedge-buying through the third quarter of 1965. Although labor and management reached agreement before the 1968 strike deadline, hedge-buying, of imports particularly, resulted in large deliveries throughout the first three quarters of 1968.

Import Transport Strikes

Consideration of dock and shipping strikes requires the addition of another independent variable to the multiple regression equation. This variable equals the percentage of man-days lost in each quarter due to major strikes in the maritime industry and appears in Table A-9.

The BLS provides data on major strikes, which they define as those involving more than 10,000 workers. The BLS also publishes the total number of workers in water transport services over the last fifteen years.[22]

The Import Demand Equation

Regressing the log of steel import volume on the five independent variables in equation (3.1), one obtains the following regression equation (with standard errors in parentheses) for quarterly data between 1963 (1) and 1968 (4):

$$\begin{aligned} M = &- \underset{(0.139)}{0.493} - \underset{(0.979)}{4.170}\,P + \underset{(0.559)}{2.307}\,Y \\ &+ \underset{(0.289)}{0.095}\,C + \underset{(0.063)}{0.176}\,L_1 - \underset{(0.426)}{0.243}\,L_2 \end{aligned} \qquad (3.2)$$

$R^2 = 0.95 \qquad D.W. = 1.73$

The coefficients of P (4.17) and Y (2.307) give values for the price and income elasticities of steel import demand. The coefficients and standard errors for C and L_2 indicate that neither capacity utilization nor transport strikes have a significant influence on import demand. However, steel labor disputes appear to significantly affect imports.

Since the elasticity estimates, in particular, are crucial to accurately assessing the impact of trade restrictions, we should stop to consider the sensitivity of the results to secular change. For instance, observed price elasticities might differ over time with the lowering of water in the United States tariff. Alternatively, previous studies suggest that elasticities probably increase after buyers become familiar with the import market. For the steel industry, in particular, new trade channels had been established during the 1959 steel strikes when United States imports rose. After discovering that they could use the cheaper foreign product, United States steel consumers may need only slight price differences to rationalize a venture into or out of the import market.[23]

To test for changes in elasticities for an expanded period, regression equations in the form of (3.2) for 1954 (1) to 1958 (4), 1959 (1) to 1963 (4), and 1954 (1) to 1968 (4) were subjected to a test for equality between regression coefficients.[24] We reject the hypothesis of equality between the five-year regression coefficients with a 99 percent confidence interval, and thereby confirm the expectation that price elasticities seem to be growing over time. American buyers do become more responsive to slight price changes as import volume grows, and the shifts in the function may be due to the learning process suggested above.

Domestic Demand and Supply Elasticities

Estimating the impact of domestic subsidies on import demand requires data on domestic price elasticities. We know little about these elasticities, especially for investment goods like steel, but Yntema made one estimate that lies between 0.1 and 0.2.[25]

Yntema also formulated the a priori relationship between domestic demand and supply elasticities and import demand elasticities[26] ($\eta_{D'}$, ϵ_D, and η respectively). He derives this relationship by defining the quantity of import demand (M) as the difference between domestic demand (D) and supply (S), at prices (P) below domestic equilibrium. We can transform the equation for import demand elasticity (employing d to denote a small change) into an expression containing domestic elasticities:

$$\begin{aligned}\eta &= (dM/M)(P/dP)\\ &= (P/dP)(dD + dS)/(D - S)\end{aligned}$$

$$= dD(P/dP)/(D - S) + dS(P/dP)/(D - S)$$

$$= DdD(P/dP)/D(D - S) + SdS(P/dP)/S(D - S)$$

$$= (D/M)\eta_D + (S/M)\epsilon_D \qquad (3.3)$$

As equation (3.3) illustrates, the import demand elasticity for steel maintains an inverse relationship with the share of imports in domestic consumption and production. Using this relationship, the import shares in the statistical appendix, our estimate of the United States import demand elasticity (4.17), and Yntema's average domestic demand elasticity estimate of 0.15, we can derive the elasiticity of domestic supply for steel:

$$\epsilon_D = [\eta - (D/M)\eta_D](M/S)$$

$$= 0.146. \qquad (3.4)$$

We must qualify this or any other statistical estimate of supply elasticities, however. The qualification, as Milton Friedman points out, arises because we find it difficult to define the average cost of a particular firm for different hypothetical outputs independently of conditions of demand due to the existence of specialized resources.[27] Even under constant cost conditions the supply curve of a monopolistic industry like steel may slope upward depending on the manner in which demand shifts outward.[28] Although competitive firms operating in the range of constant returns to scale possess infinitely elastic domestic and export supply functions, the same may not be true for the monopolist. Nevertheless, even the imperfectly competitive industry's supply depends, at least in part, on its factor costs.

These, we have noted in Chapter 1, appear to trend upward for labor and slightly downward for materials, although productivity increases moderate the effects on steel prices.[29] On the basis of this data, we might conclude that all the major steel-trading countries could expand steel production in the future without increasing their average costs.

In the short run, however, capacity utilization and the tightness of the labor market may impose limits on the supply of steel. During inflationary periods with little unemployment, the cost of attracting more labor and other resources to the steel sector, for instance, would become high. Recent complaints by United States steel producers about tight labor markets[30] illustrate the difficulties of expanding steel output when there is full employment. In contrast, when higher rates of unemployment exist, the United States and the other major steel traders could conceivably achieve a costless reallocation of labor and their domestic supply elasticities would assume correspondingly greater values. Substantial variations in capacity utilization over the years and corresponding periods of stable and rising prices also lend support to the view that domestic supply elasticities depend strongly on the industry's operating rate. Therefore the estimates in Chapter 4 will employ alternative assumptions concerning domestic supply elasticities. First, we assume that they equal infinity, so that import demand decreases by the full value of a subsidy. Second, the domestic supply elasticities will take the value of 0.146.

Foreign Import Demand Elasticities

There are numerous statistical estimates of import demand elasticities for some countries.[31] A few of these estimates apply to steel, and several more to broader commodity groups in which steel is an important item. We can find specific steel estimates, however, only for some foreign countries not considered in this study[32], or for significantly different time periods.[33] Deficiencies in country data in some cases make direct price elasticity estimates impossible. In other cases they would make the results incomparable with the United States estimates, since the latter are based on quarterly data for a precisely defined steel category. Therefore foreign import demand elasticities are inferred from readily available data and compared with independent estimates for broader commodity groups.

The Assumption of Identical Domestic Elasticities

Similarities in culture and other demographic factors, state of economic development, pattern of industrialization, technology, and steel industry structure among the ten countries considered here suggest that their domestic demand and supply schedules are also about the same. Such an assumption has been used before in studies of differences in effective tariffs, and estimates of the restrictiveness of tariffs.[34] We employ this assumption here in order to facilitate the calculation of foreign import demand elasticities, and we can compare the results with the few independent estimates available.

The Method of Calculation

We infer foreign price elasticities from the United States estimates, import share ratios, and the a priori relation between the elasticities and ratios, as shown in formula (3.3). On the assumption of equal domestic demand and supply elasticities for steel in all industrial countries, we can derive foreign elasticities (η) by dividing the domestic demand and supply elasticities by the share of imports in domestic consumption and production respectively, and summing the absolute value of their products. The appropriate import ratios for this computation were obtained from the Organization for Economic Cooperation and Development (OECD), and we present them in Table A-10 of the Appendix.

Note that domestic supply elasticities (ϵ_D) may differ markedly under different economic conditions. Since import demand depends in part on domestic supply, alternative estimates of supply elasticities will yield different import demand figures. In particular, if the industry is operating in the infinitely elastic range of the domestic supply schedule, equation 3.3 shows that import demand elasticities (η) will also be infinite. On the other hand, if $\epsilon_D = 0.146$, then the η for each country will equal the values appearing below in Table 3-1.

Other Estimates

Independent verification of the elasticities in Table 3-1 is possible for a few countries. DeVries' steel-import-demand-elasticity estimates based on information provided by United States commodity experts range from 3.14 to 4.53.[35] This study's statistical estimate for the United States falls within this range and also lies close to Kreinin's estimate of 4.71 for United States imports of manufactures.[36] Similarly, Scott calculated the British price elasticity of demand for imported manufactures and found it to equal to 6, a figure which is close to the inferred elasticity for the United Kingdom.[37]

For the most part, Houthakker and Magee's recent elasticity estimates for total imports are lower than other independent results.[38] Nevertheless, they tend to support this study's conclusion that Japan has one of the highest elasticities of import demand of the ten countries considered here.

Table 3-1
Elasticities of Import Demand for Steel with Respect to Price

Belgium	1.72
Canada	2.52
France	1.21
West Germany	1.95
Italy	1.69
Japan	44.44
Luxembourg	1.72
Netherlands	0.44
United Kingdom	4.78
United States	4.17

Source: See text.

Moreover, ranking their elasticities for the original EEC countries by size shows that the relative positions of their estimates and this study's are substantially the same: Belgium-Luxembourg and Germany have the highest elasticities, followed in order by Italy, France, and the Netherlands.

The present study's Canadian elasticity estimate of 2.52 receives the most support and this independent verification comes from several sources. Kemp calculated elasticities ranging from 1.28 to 3.97 for imports in a steel-dominated category called "materials for investment in structures." The average of these estimates comes very close to the inferred elasticity. Moreover, Kemp assigned another important steel group including equipment for producers to an elasticity of 2.61.[39]

World Demand for Steel Imports

Studying the impact of export subsidies requires a knowledge of the export demand facing the subsidizing country, and this depends partly on the total demand for imports. Consequently, it is useful to note the relationship between the elasticity of import demand for each country (η_i) and the elasticity of world demand for imports (η_w). As the following derivation illustrates η_w is the weighted average of η_i. The weights are the share of each (i_{th}) country's imports (M_i) in total world imports (M_w). Where P is the world trade price, and d denotes a small change:

$$\begin{aligned} \eta_w &= (dM_w/M_w)(P/dP) \\ &= \sum_{i=1}^{n} (dM/M_w)(P/dP) \text{ since } M_w = \sum_{i=1}^{n} M_i \\ &= \sum_{i=1}^{n} (dM_i/M_i)(P/dP)(M_i/M_w) \\ &= \sum_{i=1}^{n} \eta_i(M_i/M_w). \end{aligned} \tag{3.5}$$

Since the ten major steel-trading countries account for about 80 percent of free-world imports, the weighted average of their import demand elasticities should yield a reasonable approximation of the elasticity of world demand for steel. Therefore averaging the figures appearing in Table 3-1 and weighting them according to the import shares in the Appendix yields a value of 3.162 for η_w. Alternatively, if any $\eta_i = \infty$ (due to $\epsilon_D = \infty$), then $\eta_w = \infty$.

Export Supply Elasticities

Method of Calculation

The elasticities of steel export supply and import demand determine the distribution of the burden of a tariff or other cost imposed by trade restrictions. It also shows the quantity of exports retracted due to the imposition of trade barriers. The more inelastic is the export supply, the more the exporter absorbs the cost, and the less he decreases the quantity supplied; we consider estimates of supply elasticities here and check them against some independent observations. The estimation method relies on the precise a priori relationship between domestic demand and supply elasticities, and export supply elasticities (η_D, ϵ_D, and ϵ_x respectively). This connection follows from considering the quantity of exports (X) to be the difference between the quantity of domestic supply (S) and demand (D) at prices (P) above domestic equilibrium. Employing d to indicate a small change, the equation for export supply elasticities can be written:

$$\begin{aligned} \epsilon_x &= (dX/X)(P/dP) \\ &= [(dS + dD)/dP]P/(S - D) \\ &= (P/S)(dS/dP)[S/(S - D)] + (P/D)(dD/dP)\,[D/(S - D)] \\ &= \epsilon_D(S/X) + \eta_D(D/X). \qquad (3.6) \end{aligned}$$

Using equation (3.6), the information on the share of exports in production and consumption shown in Table A-11, and the domestic elasticities (0.150 and 0.146) discussed earlier, we can calculate export supply elasticities (ϵ_{xi}) for each of the major steel-trading countries. We see these estimates in Table 3-2. Of course, the alternative estimate of infinite domestic supply elasticities ($\epsilon_D = \infty$) would yield infinite export supply elasticities for each country. It also follows in the latter case that the world supply of steel exports would be infinitely elastic (i.e., $\epsilon_w = \infty$). Where $\epsilon_D = 0.146$, on the other hand, the relationship between ϵ_{xi} and ϵ_w is not so readily apparent. Nevertheless, the evaluation of export subsidies requires knowledge of the demand for a country's exports, which in turn depends in part on world export supply.

Therefore, we derive the relationship between ϵ_{xi} and ϵ_w, and find that ϵ_w equals the weighted average of ϵ_{xi}. The weights are the share of each (i_{th})

Table 3-2
The Price Elasticity of the Supply of Steel Exports from Each of the Major Steel-trading Countries

Belgium	0.209
Canada	3.977
France	0.831
Germany	0.817
Italy	2.431
Japan	1.831
Luxembourg	0.209
Netherlands	0.310
United Kingdom	1.849
United States	34.194

Source: See text.

country's exports (X_i) in total world exports (X_w). Where P is the world trade price and n the number of countries:

$$\begin{aligned}\epsilon_w &= (dX_w/X_w)(P/dP) \\ &= \sum_{i=1}^{n} (dX_i/X_i)(P/dP)(X_i/X_w) \text{ since } \sum_{i=1}^{n} X_i = X_w \\ &= \sum_{i=1}^{n} \epsilon_{xi}(X_i/X_w). \qquad (3.7)\end{aligned}$$

Since the ten major steel-trading countries account for about 90 percent of free-world exports, the weighted average of their export supply elasticities gives a good approximation of the world export supply elasticity. After averaging the figures in Table 3-2 with weights shown in Table A-11, one obtains a value of 1.755 for ϵ_w, where $\epsilon_D = 0.146$.

Note that ϵ_w is not the appropriate elasticity to match with each country's import demand elasticity (η_i) in order to assess the impact of trade restrictions. Instead, the elasticity of supply of exports (ϵ_i) facing each country (i) must be determined, and it is precisely related to ϵ_{w-i}. Using the standard notation of this chapter, we can define ϵ as follows:

$$\begin{aligned}\epsilon_i &= (dX_{w-i}/M_i)(P/dP) \\ &= (dX_{w-i}/X_{w-i})(P/dP)(X_{w-i}/M_i) \\ &= \epsilon_{w-i}(X_{w-i}/M_i) \qquad (3.8)\end{aligned}$$

Thus, ϵ and ϵ_w vary inversely according to the share of world exports which the ith country imports. We list these shares in the Appendix, and the elasticity of supply of exports to each country appears in Table 3-3 below.

Table 3-3
The Price Elasticity of the Supply of Steel Exports to Each of the Major Steel-trading Countries

Belgium	79.82
Canada	78.24
France	17.88
Germany	16.36
Italy	28.87
Japan	216.79
Luxembourg	149.17
Netherlands	37.21
United Kingdom	58.09
United States	9.72

Source: See text.

Other Estimates

Direct statistical estimation of the export supply elasticities is difficult mainly because of an identification problem. The extent to which our statistical estimate of the United States import demand elasticity actually measured demand and not supply depended on a stable demand function and a highly variable supply function. The same attributes prevent the same price-quantity relations from specifying an export supply elasticity.

Nevertheless, casual inspection of empirical evidence tends to support the estimate that the elasticity of supply (ϵ) to America from the rest of the world is high, at least in the long run. The index of United States import prices in the Appendix shows that prices did not climb significantly until the last few years. The average United States f.o.b. import price was $116 in the fourth quarter of 1954, compared with a 1968 price of $111. Over the same period, United States import volume increased by 2100 percent. Although we observe large quarterly price variations in these years, the trend indicates that the long-run foreign supply of exports to the United States is very elastic.

A very elastic United States supply is indicated by observing the United States export market, where export prices declined from $207.65 per ton in 1961 to $198.45 in 1965. During this five-year period, United States export volume increased from 1,990 to 2,496 thousand tons. Likewise, the response of United States producers to competition in their export markets indicates a very elastic export supply. According to a congressional study, they make few attempts to align export pricing on the substantially lower prices quoted in third markets by the European or Japanese steel producers. "In recent years, these price policies have meant that the United States producers withdrew from exporting" rather than lower price.[40]

Table 3-4
The Price Elasticity of Demand for the Steel Exports of Some Major Steel Traders

Belgium	28.23
France	38.80
Italy	127.14
Japan	23.42
Luxembourg	58.20
Netherlands	105.01
United Kingdom	63.57

Source: See text.

We may estimate ϵ to America more precisely using the relationship between it and η on the one hand, and the distribution of the burden of a tariff change on the other. The principles of incidence developed in Chapter 2 lead to the relationship below. Where t denotes the change in tariffs as a percentage of f.o.b. import price, and p denotes the percentage change in f.o.b. import price,

$$\epsilon/\eta = (t - p)/p$$

$$\epsilon = \eta(t - p)/p \tag{3.9}$$

We obtain tariff rates as well as average prices for the year preceding and following Dillon Round tariff concessions from f.o.b. unit value data provided by the Department of Commerce.[41] We then can compute the differences between prices as a percentage of their average. The resulting value for p is 0.006, and t equals 0.015. The estimate for ϵ to America derived from equation (3.9) is 6.09.

Of course, we cannot tell if t was the only influence on p, because in the absence of a comparable group of imports whose duties were not reduced, the impact of other factors affecting p cannot be measured. To the extent that this estimate of ϵ for the United States is reliable, however, it tends to substantiate the inferred estimate of Table 3-3 that the foreign supply of exports to the United States is elastic.

In addition, one independent source gives a statistical estimate of the elasticity of supply for exports of steel plate from Germany (0.27).[42] While this figure lies below the one derived here (0.817), they both show that German supply is inelastic. The same study also found ϵ_x of the United Kingdom to be fairly small but larger than the German figure—a result which conforms with the present study's elasticity estimate.

The Elasticity of Demand for Exports

The incidence of an export subsidy determines the extent to which it erects an import barrier and depends on the elasticity of export supply and the elasticity of demand facing the (i_{th}) subsidizing country's exports (η_{xi}). We have not estimated the latter elasticity, though it is quite possible to derive it from information on export shares, the elasticity of world demand for imports (η_w) and the elasticity of supply from the rest of the world (ϵ_{w-i}). Denoting the quantity of exports supplied by the rest of the world as X_{w-i}, and otherwise using a notation which should be quite familiar by now, one expresses η_{xi} in the following way:[43]

$$\begin{aligned}
\eta_{xi} &= (dX_i/X_i)(P/dP) \\
&= (P/dP)d(X_w - X_{w-i})/(X_w - X_{w-i}) \text{ since } X_w = X_i + X_{w-i} \\
&= (dX_w/dP)[P/(X_w - X_{w-i})] + (dX_{w-i}/dP)[P/(X_w - X_{w-i})] \\
&= (dX_w/X_w)(P/dP)X_w/(X_w - X_{w-i}) \\
&\quad + (dX_{w-i}/X_{w-i})(P/dP)X_{w-i}/(X_w - X_{w-i}) \\
&= \eta_w(X_w/X_i) + \epsilon_{w-i}(X_{w-i}/X_i). \qquad (3.10)
\end{aligned}$$

The export shares of each steel country appear in Table A-11, and the values for η_w and ϵ_{w-i} have been estimated. This information and equation (3.10) yield estimates of η_x for the seven major steel traders which grant export subsidies: the estimates for the case in which $\epsilon_D = 0.146$ appear in Table 3-4. Alternatively, if η_w or ϵ_{w-i} are infinite, then η_x for any country equals infinity.

Notes

1. R. Rippe, "Wages, Prices, and Imports in the American Steel Industry," *Review of Economics and Statistics* (February 1970), pp. 34-45.

2. R. Ball and K. Marwah, "The United States Demand for Imports, 1948-1958," *The Review of Economics and Statistics* (November 1967), pp. 395-401; and M. Kreinin, "Price Elasticities in International Trade," *Review of Economics and Statistics* (November 1967), pp. 510-16.

3. With imports lagged one quarter after price and income, the resulting regression equation was:

$$M = \underset{(0.062)}{-0.478} + \underset{(0.332)}{0.242}\ P + \underset{(0.280)}{5.067}\ Y - \underset{(0.128)}{0.262}\ C + \underset{(0.064)}{0.351}\ L_1 - \underset{(0.663)}{0.966}\ L_2$$

R = 0.969

4. United States Department of Commerce, *World Trade Information Service,* Part 3: Statistical Reports, 1954-1960, and *Overseas Business Reports*, 1961-1968, Washington, D. C.: United States Government Printing Office.

5. Ibid.

6. I. Kravis, and R. Lipsey, *Price Competitiveness in World Trade*, National Bureau of Economic Research, 1971, p. 5.

7. United States Department of Labor, Bureau of Labor Statistics, *Wholesale Price Index*, Washington, D. C.: United States Government Printing Office.

8. See United States Department of Labor, Bureau of Labor Statistics, *BLS Bulletin*, no. 1214 (September 1967), p. 3; no. 1257 (July 1959), p. 51; no. 1411 (June 1965), pp. 27, 90, 91, Washington, D. C.: United States Government Printing Office.

9. United States Department of Labor, Bureau of Labor Statistics, "Handbook of Methods for Surveys and Studies," *BLS Bulletin*, no. 1458, Washington, D. C.: United States Government Printing Office; 1967, p. 91, 92.

10. Joint Economic Committee, *Steel Prices, Unit Costs, Profits, and Foreign Competition* (Washington, D. C.: United States Government Printing Office, 1963), p. 244.

11. M. Bailey, "Administered Prices in the American Economy," *The Relationship of Prices to Economic Stability and Growth: Compendium of Papers Submitted by Panelists* Appearing before the Joint Economic Committee, 85th Cong., 2nd sess., March 31, 1958 (Washington, D. C.: United States Government Printing Office, 1958), pp. 91-95.

12. J. Flueck, "A Study in Validity: BLS Wholesale Price Quotations," *The Price Statistics of the Federal Government* (New York: National Bureau of Economic Research, 1961), pp. 432-33. "Labor Department Examines Consumer Prices of Steel," *The Iron Age,* April 25, 1946.

13. Kravis and Lipsey, *Price Competitiveness*.

14. H. McAlister, "Statistical Factors Affecting the Stability of The Wholesale and Consumer Price Indexes," *The Price Statistics of the*

Federal Government, New York: The National Bureau of Economic Research 1961.

15. G. Stigler, and J. Kindahl, *The Behavior of Industrial Prices* (New York: National Bureau of Economic Research, 1970), pp. 72, 74.

16. United States Department of Commerce, *Survey of Current Business and Business Statistics*, Washington, D. C.: United States Government Printing Office, various years.

17. C. Kindleberger, *International Economics*, Irwin, pp. 162-63.

18. American Iron and Steel Institute, *Annual Statistical Report*, 1958, 1959, p. 13.

19. Senate Judiciary Committee, *Administered Prices: Steel*, (Washington, D. C.: United States Government Printing Office, 1957), p. 549.

20. *The Iron Age* (September 5, 1968), p. 85.

21. *The Iron Age*, May 6, 1954, p. 199 and June 10, 1954, p. 163; April 28, 1955; January 29, 1959, p. 129; January 7, 1960, p. 371; February, 15, 1962, p. 143; April 18, 1963, p. 163; June 17, 1965, p. 97; February 29, 1968, p. 45; July 4, 1968, p. 81; and September 5, 1968, p. 85.

22. United States Department of Labor, Bureau of Labor Statistics, "Work Stoppages-Water Transportation Industry, 1927-59," *BLS Report*, no. 176, p. 7 and *BLS Bulletin*, no. 1460, p. 22, and United States Department of Labor, Bureau of Labor Statistics, "Employment and Earnings Statistics," *BLS Bulletin*, no., 132-5, p. 839.

23. Council of Economic Advisors, *Report to the President on Steel Prices*, p. 12; and Rippe, "Wages, Prices, and Imports."

24. For the method, see G. Chow, "Test of Equality between Sets of Coefficients in Two Linear Regressions," *Econometrica* 28 (July 1960), pp. 591-605. The regression equations are:

1954-1958:

$$M = \underset{(0.116)}{-0.340} - \underset{(0.411)}{1.438}\,P + \underset{(1.327)}{4.904}\,Y + \underset{(0.286)}{0.079}\,C + \underset{(0.118)}{0.049}\,L_1 - \underset{(3.513)}{1.190}\,L_2$$

$R^2 = 0.68$ D.W. = 0.71

1959-1963:

$$M = \underset{(01.75)}{-0.143} + \underset{(0.771)}{0.384}\,P + \underset{(1.277)}{4.158}\,Y - \underset{(0.184)}{0.023}\,C + \underset{(0.122)}{0.481}\,L_1 - \underset{(1.487)}{0.877}\,L_2$$

$R^2 = 0.62$ D.W. = 1.54

1954-1968

$$M = -0.453 - \underset{(0.059)}{0.932}\,P + \underset{(0.326)}{4.335}\,Y - \underset{(0.273)}{0.215}\,C + \underset{(0.121)}{0.037}\,L_1 - \underset{(0.063)}{0.828}\,L_2$$
$$(0.647)$$

$R^2 = 0.94$ D.W. = 0.95

25. Temporary National Economic Committee, "The Iron and Steel Industry: United States Steel Corporation Studies, Prices and Costs," *Investigation of Concentration of Economic Power,* Part 26, (Washington, D. C.: United States Government Printing Office, January 1940), pp. 13592-94.

26. T. Yntema, *A Mathematical Reformulation of the General Theory of International Trade* (Chicago: University of Chicago Press, 1932), p. 43-45.

27. See M. Friedman, *Price Theory: A Provisional Text* (Chicago: Aldine, rev., 1962), p. 142.

28. C. Ferguson, *Microeconomic Theory* (Homewood, Illinois: Irwin, 1966), pp. 236-38; ibid., p. 54.

29. See United Nations, Economic Commission for Europe, *International Comparisons of Labour Productivity in the Iron and Steel Industry* (New York: United Nations, 1967), p. 13; and OECD, *The Iron and Steel Industry*, various years.

30. M. Drapkin, "Steel Workers Become Much Sought After as Mills Expect Continued Strong Activity," *The Wall Street Journal* (May 21, 1969), p. 7.

31. For references to estimates for a few steel products, see H. S. Chend, "A Collection of Statistical Estimates of Elasticities in International Trade," *International Monetary Fund Staff Papers* (April, 1959), pp. 107-158.

32. See J. Polak, *An International Economic System*, Chicago: University of Chicago Press, 1953; and D. Morgan and W. Cortlett, "The Influence of Price in International Trade: A Study in Method," *Journal of the Royal Statistical Society*, Series A. CXIV Part III, 1951, pp. 307-52.

33. T. Chang, *Cyclical Movements in the Balance of Payments*, Cambridge, England, 1951.

34. B. Balassa, *Trade Liberalization among Industrial Countries* (New York: Council of Foreign Relations, 1967), p. 49; and M. E. Kreinin, *Alternative Commercial Policies: Their Effect on the American Economy*, Michigan State University, 1967.

35. For a review of estimates that have been made, see B. A. de Vries, "Price Elasticities of Demand for Individual Commodities into the United States," *International Monetary Fund Staff Papers* (April 1951), pp. 43-45.

36. Kreinin, *Alternative Commercial Policies*, p.46.

37. He calculates a very large import demand elasticity (-15) for steel in postwar years, but notes that this is the result of the unusual conditions of British steel shortages in the face of government price control. See M. Scott, *A Study of United Kingdom Imports*, Cambridge University Press, 1963, p. 184.

38. H. Houthakker and S. Magee, "Income and Price Elasticities in World Trade," *Review of Economics and Statistics* (May 1969), p. 113.

39. M. Kemp, *The Demand For Canadian Imports*, 1926-55, University of Toronto Press, 1962, Table 4.1.

40. This paragraph is based on the Senate Finance Committee's *Steel Imports*, pp. 76, 78, 126.

41. United States Department of Commerce, Bureau of the Census, *United States Commodity Exports and Imports as Related to Output*, Washington, D. C.: United States Government Printing Office, 1958 and 1963; and United States Department of Commerce, World Trade Information Service, Part 3: *Statistical Reports*, 1954-1960, and *Overseas Business Reports*, Washington, D. C.: United States Government Printing Office: 1961-1968.

42. Morgan and Corlett, "The Influence of Price," p. 351.

43. This formula is also derived in F. Shorter, "Jute Production Policies of India and Pakistan," *Indian Economic Journal* 3 (July 1955), p. 44.

4 The Restrictiveness of Tariffs and Nontariff Barriers

Using the absolute values of elasticities derived in the preceding chapter and the information on trade restrictions summarized in Chapter 2, we can estimate the value of steel imports excluded by each of the major steel-trading countries. To determine the restrictive effects of the barriers, it is convenient to divide them into five groups: government procurement policies, quantitative restrictions, subsidies, other nontariff barriers, and tariffs.

Government Procurement Policies

One method of measuring the amount of potential imports excluded by government procurement policies consists of comparisons between import/consumption ratios for the public and private sectors of each country.[1] The extent to which these ratios differ would reflect the amount of bias in government purchasing; the increase in government imports necessary to bring the ratios into equality would indicate the amount of potential trade deterred. The OECD tried to make such a comparison, but could not obtain the necessary information, since most government agencies do not specify their foreign purchases, nor report their expenditures according to product category.[2]

In the absence of data on actual government imports of steel (except for the United States Agency for International Development), we can estimate only the restrictiveness of the percentage price preferences applied to government procurement, summarized in Table 5-1 (Chapter 5). These price preferences are equivalent to ad valorem import duties on c.i.f. value including duty, an equivalency which arises from the similarity between tariffs and nontariff barriers discussed in Chapter 2. Since the trade data are valued exclusive of duty, however, we express the preferences as a percentage of c.i.f. duty-free import prices. Multiplying the percentage preferences by one plus the post-Kennedy Round (pre-Kennedy Round for Canada) tariff rate in each country yields the appropriate nontariff barrier rates. They appear in the first row of Table 4-1.

Assuming that import demand and supply elasticities from Chapter 3 are applicable to each government, we can estimate the value of steel imports excluded by the price differential. Of course, such estimates do not take into account the unquantified domestic preferences of the British, French, German, Italian, and Japanese governments and may not reflect all the bias in the purchasing policies of the Benelux and North American governments.

The duty-free value of restricted steel imports can be obtained by figuring the decrease in potential government imports resulting from imposition of the percentage price preference. We estimate potential government steel imports (V) in the absence of price preferences as follows. Assume that a government's share of total domestic steel consumption is proportional to its share of total domestic consumption of all other goods. Then the ratio of total government expenditures to GNP times the total value of steel consumed in each country yields an estimate of the value of steel purchased by each government. We can see these estimates for 1967 in Table 4-1, row 2. If each government and its corresponding economy had the same import/consumption ratio for steel in the absence of purchasing preferences, then total government imports (shown in Table 4-1, row 3) should equal total government consumption times the import ratio of the country as a whole.

If the price margin as a percentage of the c.i.f. import price is denoted by t, then in the eyes of government purchasing agents the percentage rise in the price of imports would equal dP'/P (see equation (2.2) in Chapter 2) while the actual c.i.f. import price would fall by dP/P, as in equation (2.3). Given the values for potential government imports and price preferences which appear in Table 4-1, and the appropriate elasticities from Chapter 3, we employ equations (2.6, 7, 8) to calculate alternative estimates of the value of steel imports restricted by price preferences in each of the major steel-trading countries.

We can employ equation (2.6) to estimate the restrictiveness of price preferences whenever near-capacity production occurs at home and abroad. Whenever substantial excess capacity exists only at home, equation (2.8) applies, while the case of excess capacity abroad is accounted for by equation (2.7).

Table 4-2 shows the total decline in each country's imports, as well as the reduction in each steel trader's exports. We compute the change in government imports from each foreign country on the assumption that steel from each one would account for the same share of both private and government purchases of steel imports in the absence of price preferences.

For most nontariff barriers the analysis of restrictiveness proceeds on the rather safe assumption that no trade diversion results from discrimination on the part of ths EEC and EFTA. The diversionary effects are small, if not nonexistent, because nontariff barriers usually apply to all imports irrespective of origin.[3] In the case of government price preferences, however, trade diversion results because Benelux members do not impose

Table 4-1
Steel Consumption, Imports, and Price Preferences of Governments in the Major Steel-trading Countries

	Importing Countries:						
	Belgium	*Canada*	*Luxembourg*	*Netherlands*	*United States Defense*	*AID*	*Other*
Price preferences[a]	10.720	10.670	10.720	10.720	30.836		6.522
Estimated government steel consumption[b]	50.949	340.837	1.700	64.768	2035.155	163.000	2172.137
Estimated government steel imports							
Total	19.615	39.196	.654	43.200	138.390	145.070	147.705
From Belgium		2.211	.496	12.476	12.856	13.477	13.722
Canada	1.922		.000	.108	13.590	14.246	14.505
France	3.491	1.325	.039	4.272	8.456	8.864	9.025
Germany	5.355	2.689	.061	21.233	21.603	22.645	23.057
Italy	.235	.188	.003	.682	2.878	3.017	3.072
Japan	.502	4.166	.006	.199	57.238	60.001	61.091
Luxembourg	8.999	.737		4.156	4.276	4.483	4.564
Netherlands	1.922	.016	.022		2.671	2.800	2.851
United Kingdom	.317	3.869	.004	1.171	9.936	10.416	10.605
United States	.907	21.895	.010	.376			
Total government imports subject to price preferences[c]	8.694	39.196	.136	26.568	138.390		147.705

[a]As a percentage of c.i.f. import price excluding duties and the preference.

[b]The c.i.f. value in millions of dollars.

[c]Less than row three for Benelux members by the amount of imports from other members.

Source: See text.

Table 4-2
Steel Imports Excluded by Government Purchasing Policies (C.I.F. Value in Millions of Dollars)

	Estimated Value of Excluded Imports from:										
	Belgium	*Canada*	*France*	*Germany*	*Italy*	*Japan*	*Luxem-bourg*	*Nether-lands*	*United Kingdom*	*United States*	*Total Imports Excluded*
Importing Country (2.6)											
Belgium		.301	.447	.683	.034	.072			.045	.163	1.585
Canada	.262		.190	.384	.034	.708	.087	.002	.658	5.671	10.666
Luxembourg		.000	.005	.008	.001	.001			.001	.002	.013
Netherlands		.640	.007	.272	1.143	.038			.065	.018	2.677
United States											
Defense Dept.	4.332	9.328	3.518	8.945	13.185	29.920	1.441	.936	3.087		138.390
AID	12.407	13.129	8.169	20.857	2.767	55.298	4.131	2.580	9.586		133.700
Other	1.019	2.345	.877	2.228	.421	7.611	.339	.223	1.325		22.923

(2.7)											
Belgium		.354	.643	.987	.043	.093			.058	.167	1.603
Canada	.602		.361	.731	.051	1.134	.201	.004	1.052	5.958	10.665
Luxembourg		.000	.007	.011	.001	.001			.001	.002	.025
Netherlands		.588	.005	.202	1.002	.032			.055	.018	1.253
United States											
Defense Dept.	16.531	17.475	10.873	27.778	29.418	73.600	5.498	3.435	12.776		138.390
AID	12.407	13.129	8.169	20.857	2.767	55.298	4.131	2.580	9.586		133.700
Other	03.732	3.945	2.455	6.271	.835	16.615	1.241	.775	2.884		40.171
(2.8)											
Belgium		.936	.033	.590	.080	.142			.095	3.060	68.694
Canada	.283		.249	.502	.064	1.184	.094	.002	1.107	21.895	39.196
Luxembourg		.000	.008	.011	.001	.002			.001	.034	.136
Netherlands		.053	.796	3.928	.231	.056			.332	1.269	26.568
United States											
Defense Dept.	4.532	15.705	4.102	10.415	2.378	39.964	1.507	.999	6.977		138.390
AID	12.407	13.129	8.169	20.857	2.767	55.298	4.131	2.580	9.586		133.700
Other	1.066	4.449	1.042	2.642	.653	10.764	.355	.239	1.880		147.705

Source: See text.

them on imports from member countries. Therefore the estimated pre-preference amount of government imports subject to this official price discrimination (shown in the last row of Table 4-1 in Belgium, Luxembourg, and the Netherlands) does not include estimated imports from other Benelux countries. Moreover, the amount of steel restricted by price preferences (shown in Table 4-2) in intra-Benelux trade is zero. The same can be said for tariffs in intra-EEC trade.

We did not subject the purchasing policies of AID to the preceding analysis because they do not state a given preferential margin. Nevertheless, AID now follows purchasing policies favoring United States steel producers. As a result, the proportion of AID steel purchases made in the United States rose from 11 percent in 1960 to 93 percent in 1965.[4] At the same time the proportion of AID steel purchased from other developed countries fell from 76 percent to less than one-half of a percent.[5] We can assess the direct restrictive effect of this biased procurement on steel imports by assuming that the same percentage of 1960 AID purchases would come from domestic steel producers if AID were to return to competitive procurement and applying the difference to AID purchases of steel, which amounted to $163 million annually in the late 1960s.[6] Foreign purchases by AID would increase by 82 percentage points on that assumption. Thus we estimate the value of steel imports excluded by AID purchasing policies to be $133.7 million.

Summarizing the results in Table 4-2, we can see that because the Benelux governments import a large share of their foreign purchases from fellow members, the volume of steel restricted by each is far less than that excluded by Canadian preferences, even though their price differentials are the same. In contrast, the large price preferences applied by the United States Defense Department, the large bias in AID purchasing, and the large volume of potential federal imports, combine to make the value of excluded United States steel imports range from almost $300 to over $400 million. Indeed, Defense Department price preferences cause that agency to forego purchase of any foreign steel.

Quantitative Restrictions

France

The French government imposed a quota of 59,000 tons per month for the last half of 1968. We can compute its restrictive effect by comparing import values with and without the quota.

In Chapter 2, we noted that the quantity of imports kept out by the quota included the amount of growth which would have occurred in its absence. We can measure this by assuming that the volume of imports is distributed evenly over the year, and that the French income elasticity of demand for steel imports equals 1.66.[7] According to these assumptions, the 4.5 percent increase in France's real GNP between 1967 and 1968 would have caused steel import demand to rise by $dM/M'' = 7.47$ percent at a given price level. We calculate potential import values with methods developed in Chapter 2, prequota import values from sources in Table 4-3, and import demand and supply elasticities from Chapter 3.

We obtain the values of excluded steel in Table 4-4 by subtracting values subject to the quota (Table 4-3, row 3) from potential import values estimated for alternative conditions. As Table 4-4 illustrates, the French quota allowed imports from many countries to increase as fast or faster than our model predicts, so that its restrictive effect was nil. Belgium, Germany, and Luxembourg bore the heaviest loss of sales. In total, the quota kept 1968 imports $2 million below 1967 levels, and excluded about $52 million worth of steel in the presence of demand growth. Thus the value of 1968 French imports could have grown almost 8 percent larger without the quota.

United States

Chapter 5 also describes "voluntary" limitations on exports to the United States. We can indicate the restrictive effect of the first of these quotas by comparing actual imports with estimates of what 1969 United States steel imports would have been in the absence of the "voluntary" limit of 14 million tons. United States GNP equalled $724.7 billion in 1969 (in terms of 1958 prices), approximately 2.6 percent over the 1968 level.[8] Given the income elasticity of demand for steel of 2.3 (computed in Chapter 3) and assuming constant relative prices, the volume of United States steel imports in 1969 should have increased by 5.98 percent to 18.8 million tons. Thus this quantitative restriction on United States steel imports in 1969 appears to restrict 4.8 million tons, a 25.5 percent decrease in import demand.

Following the same procedure used for France, we estimate the value of United States steel imports excluded by the limitation and the results appear in Table 4-5. In contrast to the results for France, we find the first United States quota keeping out between one-quarter and one-third billion dollars of steel imports in 1969. Doing away with the limitation would have increased the value of United States imports by roughly 12 to 18 percent.

Table 4-3
Value of 1968 Imports of Iron and Steel Mill Products[a] (C.I.F. in Millions of Dollars)

	Imports from:										
Importing Country	*Belgium*[c]	*Canada*	*France*	*Germany*	*Italy*	*Japan*	*Luxembourg*[d]	*Netherlands*	*United Kingdom*	*United States*	*Total Imports*
Belgium[c]		.296	53.695	82.200	3.532	7.718	138.188	29.513	4.872	13.925	321.199
Canada	11.855		10.074	20.456	1.451	31.698	4.961	.149	29.441	166.491	298.027
France	183.049	1.043		282.548	45.522	4.615	76.595	23.881	16.012	7.879	678.104
Germany	264.580	.670	218.189		68.459	9.467	110.711	90.877	16.740	5.896	987.916
Italy	36.025	3.922	74.949	108.590		11.408	15.074	13.574	10.834	9.210	348.250
Japan	.030	.009	.036	1.902	.021		.012	.019	1.088	5.645	19.776
Luxembourg[b]	52.910	.023	4.228	6.472	.278	.608		2.325	.384	1.096	69.755
Netherlands	86.873	1.105	42.196	209.728	6.747	1.975	36.351		11.581	3.738	426.647
United Kingdom	13.775	12.599	17.382	39.222	17.576	12.302	5.764	26.778		17.329	299.959
United States[e]	128.583	192.807	120.059	306.256	40.978	811.427	53.804	37.924	140.970		1961.837

[a]SITC 67 less SITC 671.
[b]Estimated from BLEU data.
[c]BLEU figures from OECD less estimated Luxembourg imports or exports.
[d]Estimated from Luxembourg/BLEU export ratio.
[e]F.o.b.
Source: OECD, *Trade Statistics*, Series C.

Table 4-4
Steel Imports Excluded by the French Quota (C.I.F. Value in Millions of Dollars)

Country of Origin	$\eta = \infty$	$\epsilon = \infty$	$\eta, e < \infty$
Belgium	3.110	17.016	15.613
Canada	.145	.234	.214
Germany	10.987	32.914	31.706
Italy	.000	.000	.000
Japan	.000	.000	.000
Luxembourg	3.188	9.148	8.591
Netherlands	.000	.000	.000
United Kingdom	.000	.000	.000
United States	.000	.000	.000
Total excluded	2.043	52.850	52.307

Source: See text.

Table 4-5
Steel Imports Excluded by the 1969 "Voluntary" Export Limits (F.O.B. Value in Millions of Dollars)

Country of Origin	$\eta = \infty$	$\epsilon = \infty$	$\eta, \epsilon < \infty$
Belgium	13.080	20.769	15.228
Canada	45.686	57.216	52.796
France	.000	6.792	2.279
Germany	94.560	112.874	101.325
Italy	7.951	10.401	9.237
Japan	47.580	96.103	70.676
Luxembourg	4.303	7.520	5.112
Netherlands	.000	1.399	.000
United Kingdom	26.398	34.828	30.439
Total excluded	238.237	355.555	329.205

Source: See text.

The American quota is not as restrictive as tonnage comparisons would indicate for several reasons. First, nonsignatory countries have increased their exports to the United States. Second, signatory countries have compensated for the tonnage limits by shipping higher-priced steel products. Third, 1969 American steel import volume fell 1.7 percent short of the quota because an unexpected increase in European steel demand caused European steel producers to not ship all they had planned. This foreign steel boom persisted through 1970 and resulted in an import volume well below the quota. Fourth, in the view of European and Japanese producers, the American surcharge of 1971 negated the agreement, and as a result American imports exceeded the quota by 2.9 million tons in that year.

The latest agreement limiting shipments from the major steel producers (except Canada) to 14.5 million tons in 1972, overcomes one of the preceding problems by specifying allowable tonnages of certain steel products. With nonsignatory shipments included, 1972 imports amount to 17.3 million tons. Using the American income elasticity as a predictor, potential 1972 imports could have been 22.5 million tons at 1968 prices. Thus, in the case of infinite elasticities of import supply, the quota excludes approximately 5.2 million tons of steel with a 1968 dollar value of $577.2 million. More detailed estimates must await complete 1972 data.

Subsidies

Domestic Subsidies

Chapter 7 describes quantifiable subsidies for steel in eight of the ten major steel-trading countries. The various ad valorem rates of subsidization (b) appear in the first column of Table 4-6, but we cannot assess their restrictive effect on imports until we convert them to ad valorem nontariff barrier rates, $b\epsilon_D/(\eta_D + \epsilon_D)$, as shown in Chapter 2. Because of alternative estimates concerning the elasticity of domestic supply, however, this calculation yields different values for the nontariff barriers created by domestic subsidies. For example, whenever $\eta_D = \infty$, a domestic subsidy creates no barrier to imports. If $\epsilon_D = \infty$, the nontariff barrier and subsidization rates are equal; but the latter exceeds the former if ϵ_D is less than infinite. The results of the computations appear in the second and third columns of Table 4-6.

Employing equations (3.13, 14, 15), and values from Table 4-3, we can estimate the increase in duty-free value arising from removal of the subsidies under alternative conditions. If $\epsilon_D = \infty$ and $\epsilon_D < \infty$ for the importing country, and $\epsilon < \infty$, then we apply equation (2.15) to nontariff barrier rates in the second column of Table 4-6. If domestic elasticities are less than infinite, then equations (2.13) and (2.14) apply (in the respective cases of $\epsilon < \infty$ and $\epsilon = \infty$) to nontariff barrier rates in the third column.[9] The results appear in Table 4-7.

Despite the fact that half of the other subsidizers import more than the United Kingdom, the latter country excludes more steel than the rest put together, whenever capacity operations prevail at home and abroad. This large value can be attributed to the size of the subsidy and the high British

Table 4-6
Nontariff Barrier Rates Created by Domestic Subsidies

Subsidizing Country	*Subsidization Rate*	*Nontariff Barrier Rates*[a] $\epsilon_D = \infty$	$\eta_D, \epsilon_D < \infty$
Belgium	0.013	0.014	0.007
Canada	0.034	0.036	0.018
France	0.010	0.011	0.005
Germany	0.012	0.013	0.007
Italy	0.008	0.008	0.004
Luxembourg	0.013	0.014	0.007
Netherlands	0.013	0.014	0.007
United Kingdom[b]	0.086	0.093	0.090

[a]Expressed as a percentage of c.i.f. price excluding duty.

[b]Eight percent is fully passed on to steel buyers; see Chapter 7.

elasticity of import demand for steel. It is also interesting to note that the British subsidy excludes an amount equal to over a third of its present 1968 imports, but a smaller proportion of its imports from each of the major steel-trading countries. This difference can be attributed to the size of the total export supply elasticity facing Britain relative to the small elasticity of the supply of exports from the individual steel traders.

Export Subsidies

Export subsidies generally create nontariff barriers equal in amount to that part of the export subsidy passed forward to buyers. It follows from Chapter 2 that whenever the subsidizing country has substantial excess capacity so that its $\epsilon_D = \epsilon_X = \infty$ the rate of export subsidization equals the nontariff barrier rate (t). We can measure the restrictiveness of the barrier in this case with equation (2.15) and rates from column one of Table 4-8. In contrast, if $\epsilon_D = \infty$, for any other country (implying that $\eta_X = \infty$), the subsidy restricts no imports. If ϵ_D is everywhere less than infinite, the barrier lies below the subsidization rate by $\epsilon_X/(\epsilon_X + \eta_X)$ as shown in the second column of Table 4-8 and equation (2.13) applies.

Table 4-9 shows alternative estimates of restrictiveness for subsidies as well as the Japanese penalty for failure to export. In Chapter 5, we established that this Japanese export incentive scheme does force the volume of steel exports to rise by 10 percent in much the same way they would behave with a subsidy. The consequent reduction in export and domestic prices

Table 4-7
Steel Imports Excluded by Domestic Subsidies (Millions of Dollars)

	Estimated Value of Excluded Imports from:										
Importing Country	*Belgium*	*Canada*	*France*	*Germany*	*Italy*	*Japan*	*Luxem-bourg*	*Nether-lands*	*United Kingdom*	*United States*	*Total Imports Excluded*
(2.13)											
Belgium		.003	.471	.719	.004	.075	1.059	.233	.048	.164	3.831
Canada	1.236		.247	.500	.045	.921	.099	.003	.557	6.904	12.614
France	1.006	.006			.275	.028	.421	.133	.096	.050	4.278
Germany	1.926	.007	1.866		.694	.092	.806	.685	.163	.074	12.021
Italy	.172	.026	.408	.590		.069	.072	.067	.065	.067	2.532
Luxembourg	.405	.000	.037	.113	.003	.006		.018	.003	.013	.835
Netherlands	.504	.004	.189	.940	.025	.008	.211		.044	.011	1.332
United Kingdom	1.452	3.061	2.491	5.591	3.635	2.302	.608	3.007		6.301	112.668

(2.14)											
Belgium		.003	.646	.989	.043	.093	.166	.355	.059	.167	3.863
Canada	.510		.433	.880	.063	1.364	.213	.006	1.267	7.165	12.825
France	1.166	.007		1.800	.290	.029	.488	.152	.102	.050	4.320
Germany	3.385	.008	2.792		.876	.121	1.416	1.163	.214	.075	12.641
Italy	.268	.029	.557	.807		.085	.112	.101	.080	.068	2.587
Luxembourg	.636	.000	.051	.078	.004	.007		.028	.004	.013	.839
Netherlands	.267	.003	.130	.645	.021	.006	.112		.036	.012	1.312
United Kingdom	5.437	4.957	6.860	15.480	6.937	4.855	2.275	10.569		6.839	118.387
(2.15)											
Belgium		.021	1.417	2.153	.176	.317	2.387	.553	.201		380.486
Canada	.529		.697	1.403	.192	3.441	.222	.007	3.216	230.328	929.412
France	2.368	.056		5.530	1.692	.141	.991	.335	.494	3.024	139.525
Germany	4.283	.045	5.394		3.195	.364	1.792	1.597	.648	2.845	234.900
Italy	.369	.167	1.168	1.679		.276	.154	.151	.264	2.785	89.365
Luxembourg	.914	.002	.111	.169	.014	.025		.044	.016	.565	153.561
Netherlands	1.500	.080	1.114	5.492	.337	.081	.628		.479	1.927	238.839
United Kingdom	1.574	6.245	3.085	6.905	5.978	3.434	.659	3.334		61.843	1799.091

Source: See text.

(dP), moreover, depends on the elasticity of demand (η_X) for Japan's exports (X). We can write η_X as follows:

$$\eta_X = (dX/X)P/dP$$

$$\eta_X = (0.1)\ P/dP$$

$$dP/P = (0.1)/\eta_X$$

If $\eta_X = \infty$, the export incentive will not create any restriction. However, if η_X for Japan equals 23.42 as estimated in Chapter 3, the percentage reduction in price, equivalent to the nontariff barrier rate created by this scheme, equals 0.43 percent of c.i.f. duty-paid value. Adjusting this to a percentage of c.i.f. value excluding tariffs, we add it to the rate of the other Japanese export subsidy shown in Table 2-7. Thus Table 4-8, column two, shows the total rate of Japanese export subsidies and penalties.

The striking feature of the figures in Table 4-9 is the degree to which alternative conditions affect the value of steel imports restricted by export subsidies. In Italy, for instance, only 4 million dollars is excluded whenever ϵ_D is everywhere less than infinite, but $3835 million when the Italian domestic supply and import demand are infinitely elastic. Since the latter value is over ten times total Italian imports, the second set of estimates in Table 4-9 must be used with caution. Such large changes may be justified when one considers the Italian subsidization rate of 29 percent, but the smaller values computed on the basis of near-capacity operations in both domestic and foreign steel industries look more realistic.[10]

Other Nontariff Barriers

We total the remaining quantified nontariff barriers and express them as a percentage of the duty-free import price in Table 4-10. We determine restrictiveness with the same equations and import data employed in previous sections. Alternative estimates of steel imports excluded by this set of nontariff barriers appear in Table 4-11. There we can see that the United States appears to keep out far more steel than the others, except in the case of infinite demand elasticities (2.15). This occurs despite the fact that American rates fall below those of four other countries.

These calculations, based largely on data applicable to 1968, do not include the restrictiveness of the temporary United States surcharge of 1971. In the absence of the "voluntary" quota, the income elasticity of American demand tells us that import volume for the last 4.5 months of 1971 valued at 1968 prices would have equalled $821.2 million. Under

Table 4-8
Nontariff Barrier Rates Created by Export Subsidies (Percentage of C.I.F. Value)

Subsidizing Country	*Subsidy Rate*[a]	*Nontariff Rate*
Belgium	7.56	0.07
France	9.39	0.25
Italy	28.83	0.70
Japan	1.20	0.52
Luxembourg	1.08	0.00[b]
Netherlands	6.01	0.02
United Kingdom	3.24	0.11

[a]Equals the nontariff barrier rate whenever $\epsilon_D = \epsilon_X = \infty$, except in the case of Japan, where 0.0163 is the nontariff barrier rate including penalty.
[b]0.0045 percent.
Source: See text.

conditions of excess capacity abroad, we can see from equation (2.7) that the surcharge-free value of excluded steel would have amounted to $342.4 million. Of course, to the extent that the quota limited imports more than the surcharge, the effects of the latter were swamped by those of the quantitative measure.

The Total Amount of Steel Restricted by Nontariff Barriers

Not all the estimates of excluded steel are strictly additive, since some nontariff restrictions were applied in years other than 1968 and some simply add water to other barriers. Nevertheless, we can obtain a rough approximation of the total imports excluded by all countries through summing the 1968 (1969 for the American quota) estimates. Hereafter, discussion about excluded import estimates concentrates on the case where elasticities are everywhere less than infinite.

Referring to the totals in Table 4-12, we can see that the United States reduces its steel imports by over 1 billion dollars with its quantified nontariff barriers. Britain and Canada, the only other countries excluding over $100 million, do not even run a close second in this race. Luxembourg, because of its small size, keeps out the least steel with its quantified barriers. The amount of United States steel exports excluded by the quantified restrictions of all major foreign steel-trading countries is less than one-tenth of the amount of United States imports restricted. However, this comparison cannot be taken as an unambiguous indication that the United States steel industry has a competitive advantage attributable to the use of

Table 4-9
Steel Imports Excluded by Export Subsidies (Millions of Dollars)

Importing Country	Belgium	Canada	France	Germany	Italy	Japan	Luxem-bourg	Nether-lands	United Kingdom	United States	Total Imports Excluded
	Estimated Value of Excluded Imports from:										
(2.13)											
Belgium		0.000	0.046	0.071	0.004	0.007	0.104	0.023	0.005	0.016	0.383
France	0.472	0.003		0.766	0.130	0.013	0.197	0.062	0.045	0.024	2.024
Italy	0.271	0.041	0.644	0.931		0.108	0.114	0.105	0.103	0.106	4.002
Japan	0.000	0.000	0.000	0.018	0.000		0.000	0.000	0.015	0.584	3.797
Luxembourg	0.000	0.000	0.000	0.000	0.000	0.000		0.000	0.000	0.000	0.000
Netherlands	0.014	0.000	0.005	0.027	0.001	0.000	0.006		0.001	0.000	0.038
United Kingdom	0.017	0.037	0.030	0.067	0.044	0.028	0.007	0.036		0.082	1.481
(2.15)											
Belgium		0.118	7.688	11.675	0.965	1.733	12.796	2.975	1.101	39.771	2109.054
France	21.118	0.524		50.243	15.642	1.301	8.837	3.003	4.545	28.413	1309.069
Italy	13.182	6.924	44.741	64.258		11.047	5.516	5.476	10.564	119.624	3834.618
Japan	0.000	0.001	0.001	0.042	0.001		0.000	0.000	0.037	2.412	52.302
Luxembourg	0.692	0.001	0.084	0.128	0.010	0.019		0.033	0.012	0.042	114.345
Netherlands	6.378	0.346	4.770	23.521	4.450	0.349	2.669		2.060	8.368	1037.104
United Kingdom	0.543	2.078	1.046	2.343	1.999	1.152	0.227	1.145		20.382	592.568

Source: See text.

Table 4-10
Rates of Other Nontariff Barriers[a] (Percentage of C.I.F. Duty-Free Value)

Belgium	10.72
Canada	16.63[b]
France	3.88
Germany	3.61
Italy	8.49
Japan	4.25
Luxembourg	6.85
Netherlands	9.82
United Kingdom	5.72
United States	7.18[c]

[a]Includes indirect border taxes, customs fees, Luxembourg's import surcharge, prior deposit requirements, documentary expenses, and freight rate discrimination.

[b]12.96 applies to only 80 percent of Canadian imports.

[c]F.o.b.

Source: See text.

nontariff barriers. For one thing, we have yet to estimate the impact of tariffs. For another, the only barriers compared here are the few that were amenable to quantification, and the number quantified may bear no relation to the total number of barriers erected. As illustrated in Table 4-13, Canada imposes the fewest nontariff restrictions, but not the smallest number of quantified ones.

Nevertheless, there are only two major steel-trading countries which raise more nontariff restrictions than the United States. Although the foreign nations may exclude relatively more steel with their nonquantifiable restraints, this cannot be determined with any precision. The large share of world steel imports accounted for by the United States tends to support the view that even if foreigners excluded half again as much, the absolute amount would still be smaller than the value of steel excluded by the United States.

From a relative standpoint, however, we do not find United States barriers among the most restrictive. Table 4-14 shows that the Japanese have clearly made the largest percentage reductions in import volume with nontariff barriers. The United States, United Kingdom, and Canada occupy middle ground, while the Europeans stand the closest to free trade.

Tariff and Nontariff Restrictions

Nominal Rates

Another interesting comparison concerns the relative importance of tariff

Table 4-11
Steel Imports Excluded by Other Nontariff Barriers (Millions of Dollars)

	Estimated Value of Excluded Imports from:										
Importing Country	*Belgium*	*Canada*	*France*	*Germany*	*Italy*	*Japan*	*Luxem-bourg*	*Nether-lands*	*United Kingdom*	*United States*	*Total Imports Excluded*
(2.13)											
Belgium		.045	7.084	10.824	.521	1.109	16.044	3.528	.701	2.293	53.233
Canada	1.881		1.971	3.990	.347	7.219	.787	.024	6.717	49.637	90.034
France	7.322	.046		11.828	1.981	.199	3.064	.966	.691	.355	30.450
Germany	10.452	.039	10.118		3.741	.496	4.373	3.717	.879	.392	64.100
Italy	3.283	.473	7.743	11.197		1.285	1.374	1.273	1.221	1.197	45.154
Japan	.002	.002	.003	.147	.003		.001	.001	.001	4.793	30.288
Luxembourg	3.919	.002	.357	.545	.027	.056		.177	.036	.119	7.666
Netherlands	6.854	.050	2.504	12.492	.325	.100	2.868		.584	.150	7.071
United Kingdom	.919	1.948	1.571	3.528	2.302	1.455	.385	1.902		4.100	73.510
United States	10.718	35.038	13.355	33.901	6.407	115.695	4.485	3.356	20.164		439.677

(2.14)											
Belgium		.049	8.492	13.689	.588	1.285	23.013	4.915	.811	2.319	53.490
Canada	3.619		3.075	6.245	.443	9.677	1.515	.046	8.988	50.829	91.563
France	8.273	.047		12.770	2.057	.209	3.462	1.079	.724	.356	30.647
Germany	17.976	.046	14.824		4.651	.643	7.522	6.174	1.137	.401	67.121
Italy	4.739	.516	9.858	14.283		1.501	1.983	1.785	1.425	1.211	45.807
Japan	.054	.016	.065	3.446	.038		.022	.034	1.971	10.227	35.828
Luxembourg	5.834	.003	.466	.714	.031	.067		.256	.042	.121	7.692
Netherlands	3.418	.043	1.660	8.252	.265	.078	1.430		.456	.147	16.786
United Kingdom	3.563	3.248	4.495	10.144	4.546	3.182	.149	6.925		4.482	77.576
United States	35.920	53.860	33.538	85.552	11.447	226.671	15.030	10.594	39.380		548.037
(2.15)											
Belgium		.171	11.052	16.783	1.398	2.505	18.242	4.250	1.591	58.009	3077.466
Canada	2.067		2.776	5.591	.777	13.896	.865	.029	3.250	950.043	1191.339
France	8.644	.208		20.267	6.227	.520	3.617	1.225	1.815	11.165	514.994
Germany	11.620	.124	14.658		8.696	.990	4.862	4.334	1.762	7.754	640.186
Italy	3.730	1.759	12.026	17.285		2.874	1.561	1.531	2.748	29.600	949.566
Japan	.002	.002	.003	.150	.003		.001	.001	.135	8.792	190.791
Luxembourg	4.434	.008	.547	.830	.069	.123		.212	.078	2.818	766.369
Netherlands	10.489	.582	7.925	39.074	2.431	.584	4.389		3.447	14.151	1753.966
United Kingdom	.962	3.739	1.868	4.181	3.589	2.066	.403	2.034		36.824	1070.856
United States	11.300	72.852	16.298	41.244	10.608	172.595	4.728	3.628	30.180		1608.324

Table 4-12
Steel Imports Excluded by All Nontariff Barriers (Millions of Dollars)

	Estimated Value of Excluded Imports from:										
Importing Country	*Belgium*	*Canada*	*France*	*Germany*	*Italy*	*Japan*	*Luxem-bourg*	*Nether-lands*	*United Kingdom*	*United States*	*Total Imports Excluded*
(2.13)											
Belgium		.349	8.040	12.285	.630	1.256	17.230	3.798	.793	2.572	57.433
Canada	2.742		2.784	5.637	.485	10.121	1.125	.035	9.415	67.427	122.240
France	24.412	.269		45.879	2.359	.238	12.273	1.160	.824	.429	88.270
Germany	12.395	.046	11.988		4.407	.586	5.187	4.409	1.038	.455	74.642
Italy	3.731	.531	8.784	12.704		1.452	1.562	1.447	1.380	1.334	50.323
Japan	.002	.002	.003	.165	.003		.001	.001	.146	5.381	33.880
Luxembourg	4.329	.002	.399	.538	.031	.063		.196	.040	.131	8.287
Netherlands	7.326	.692	2.665	13.536	1.486	.144	3.066		.682	.175	20.612
United Kingdom	2.409	5.028	4.155	9.327	6.020	3.826	1.008	5.000		9.969	177.478
United States	43.714	112.636	28.198	167.256	32.017	279.200	15.508	7.095	64.601		1063.895

(2.14)											
Belgium		.406	9.781	15.665	.674	1.471	23.179	5.270	.928	2.653	58.856
Canada	4.731		3.869	7.856	.557	12.175	1.929	.056	11.307	63.952	115.053
France	26.455	.288		47.484	2.347	.238	13.098	1.231	.826	.406	87.817
Germany	21.361	.054	17.616		5.527	.764	8.938	7.337	1.351	.476	79.762
Italy	5.007	.545	10.415	15.090		1.586	2.095	1.886	1.505	1.279	49.394
Japan	.054	.016	.065	3.446	.038		.022	.034	1.971	10.227	35.828
Luxembourg	6.470	.003	.052	.803	.036	.075		.284	.047	.136	8.556
Netherlands	3.685	.634	1.795	9.099	1.288	.116	1.542		.547	.177	19.351
United Kingdom	9.000	8.205	11.355	25.624	11.483	8.037	3.766	17.494		11.321	195.963
United States	89.359	145.625	61.827	253.332	54.868	468.287	33.420	18.783	99.454		1215.853
(2.15)											
Belgium		1.280	21.435	33.075	2.858	5.090	34.230	8.034	1.701	62.099	3202.162
Canada	3.332		4.473	9.004	1.278	22.736	1.370	.044	21.262	1539.365	6176.705
France	35.763	.989		90.179	25.072	2.078	16.851	4.664	7.257	46.281	2131.206
Germany	16.022	.175	20.445		12.252	1.392	6.705	5.992	2.477	11.035	1010.053
Italy	17.754	9.829	60.848	88.795		15.510	7.429	7.423	14.834	171.796	5505.216
Japan	.002	.003	.004	.193	.004		.001	.001	.175	11.401	247.466
Luxembourg	.113	.012	.773	1.173	.098	.176		.294	.111	4.083	1102.617
Netherlands	18.766	1.159	15.375	75.774	4.809	1.149	7.853		6.787	28.519	3404.704
United Kingdom	3.137	13.079	6.293	14.082	12.436	7.110	1.313	6.682		131.121	3817.493
United States	42.385	150.281	29.611	169.718	24.357	326.201	15.024	7.446	75.021		2266.355

Source: See text.

Table 4-13
The Distribution of Nontariff Barriers

Importing Country	*Total Number of Barriers*	*Number of Quantified Barriers*
Belgium	16	5
Canada	10	4
France	21	6
Germany	15	2
Italy	16	6
Japan	18	5
Luxembourg	15	5
Netherlands	15	5
United Kingdom	17	5
United States	17	4

Source: Tables 2-6 and 2-7.

Table 4-14
Steel Excluded by Quantified Nontariff Barriers (Percentage of 1968 Imports)

Belgium	18
Canada	41
France	13
Germany	8
Italy	15
Japan	171
Luxembourg	12
Netherlands	5
United Kingdom	59
United States	54

and nontariff barriers confronting steel. We compare them by first using equations (2.13, 14, 15) to calculate the value of steel imports excluded by post-Kennedy Round duties (pre-Kennedy Round for Canada) listed in Table 2-1. We can check the results of this calculation in Table 4-15. The tariff rates are similar, with a 2.3 percentage point difference between the highest and lowest duties in the ten countries considered here. Nevertheless, international differences in elasticities, import volume, and preferential treatment accorded intra-EEC trade result in wide disparities in the value of steel imports excluded by tariffs. The Americans exclude approximately five times more than the British, and the United Kingdom excludes at least twice as much steel as any other steel trader.

Comparing Tables 4-12 and 4-15, we can see that quantified nontariff barriers in every country save Japan exclude far more steel than tariffs. Indeed, nontariff barriers pose the only impediments to intra-EEC trade.

The United States excludes nearly equal shares with its tariffs and quantified nontariff barriers. The fact that Table 4-12 refers only to quantified nontariff barriers also serves to strengthen our impression of nontariff barriers as the predominant type of trade restriction. However, these estimates do not tell us whether nontariff barriers yield more effective protection.

Effective Rates of Protection

As we noted in Chapter 2, estimates involving nominal rates of tariffs and nontariff barriers are based on the assumption of constant restrictions on steel inputs. To consider the sensitivity of the results to changes of input barriers, we employ nominal and effective tariff rates calculated by Balassa and reviewed in Chapter 2. In addition, weighting United States tariffs[11] on various materials used in blast furnaces, steel works, rolling and finishing mills by the appropriate input coefficients,[12] yields an estimate of 0.037 for t_i. Using these facts and equation (2.1), one finds that the value of a_{ij} implicit in Balassa's results is about 0.9. Given this value, it follows that when $t_i = 0$, then $z_j = 0.77$, and the present study's estimates of excluded imports are consistent with this effective tariff rate. However, when t_i for the United States assumes its actual value of 0.037, z_j falls by approximately half to 0.377. If the same relationships between nominal and effective rates hold true for post-Kennedy Round duties, then it may be reasonable to say that the estimates of excluded imports in Table 4-15 should be halved. If this is the case, one would find the United States excluding one-quarter rather than one-half billion dollars of steel with its tariffs.

Since Balassa employs standardized input coefficients, we can also determine the sensitivity of excluded import estimates to simultaneous changes of input and output duties in major steel-trading countries besides the United States. Setting t_i equal to zero, one obtains values of z_j for each country consistent with Table 4-15 estimates based on changes in nominal output tariffs only. The various z_j under that assumption are 77 percent for the United States, 120 percent for the United Kingdom, 76 percent for Common Market countries, and 140 percent for Japan.

Contrasting the extreme values of z_j with the actual z_j computed by Balassa, we find that the ratios of the latter to the former are all less than one. Multiplying these ratios by Table 4-15 estimates indicates the reaction of imports to an across-the-board elimination of tariffs. Employing this procedure, we find that the change in United Kingdom steel imports would be 42 percent of its original value, or $47.320 million. The value of excluded imports for the EEC would be 26 percent of its original value. Therefore,

Table 4-15
Steel Imports Excluded by Tariffs (Millions of Dollars)

Importing Country	*Estimated Value of Excluded Imports from:*										
	Belgium	*Canada*	*France*	*Germany*	*Italy*	*Japan*	*Luxem-bourg*	*Nether-lands*	*United Kingdom*	*United States*	*Total Imports Excluded*
(2.13)											
Belgium		.003				.751			.474	1.586	1.845
Canada	.892		.935	1.892	.168	3.469	.373	.012	3.229	25.470	46.452
France		13.586				.365			1.267	.639	5.496
Germany		.077				.984			1.742	.757	29.855
Italy		.405				1.099			1.045	1.032	11.287
Japan	.003	.003	.005	.255	.005		.001	.001	.226	8.253	51.020
Luxembourg		.005				.059			.037	.125	.402
Netherlands		.037				.074			.435	.112	1.282
United Kingdom	1.452	3.061	2.491	5.591	3.635	2.302	.608	3.007		6.301	112.668
United States	13.009	42.390	16.230	41.200	7.770	140.436	5.443	4.075	24.476		529.337

(2.14)											
Belgium		.034				.892			.563	1.609	1.855
Canada	1.876		1.594	3.237	.230	5.016	.785	.024	4.659	26.345	47.159
France		.085				.375			1.301	.640	5.511
Germany		.088				1.240			2.192	.772	31.053
Italy		.445				1.295			1.230	1.045	11.463
Japan	.091	.027	.109	5.751	.006		.036	.057	3.289	17.067	59.791
Luxembourg		.033				.070			.044	.127	.403
Netherlands		.033				.058			.342	.110	1.261
United Kingdom	5.437	4.957	6.860	15.480	6.937	4.855	2.275	10.569		6.839	118.387
United States	42.915	64.350	40.070	102.214	13.677	270.816	17.957	12.657	47.049		654.770
(2.15)											
Belgium		.112				1.646			1.046	37.754	100.099
Canada	.971		1.273	2.565	3.494	6.273	.407	.013	5.864	418.141	1686.922
France		.395				.984			3.438	21.362	98.464
Germany		.254				2.019			3.594	15.985	316.464
Italy		1.486				2.434			2.326	24.970	232.313
Japan	.003	.003	.005	.261	.006		.001	.002	.237	15.532	337.259
Luxembourg		.009				.130			.082	2.971	40.408
Netherlands		.419				.421			2.487	10.135	125.606
United Kingdom	1.522	6.030	2.981	6.673	5.773	3.316	.637	3.224		59.689	1736.351
United States	13.728	89.289	19.880	050.306	12.986	211.097	5.744	4.411	36.914		1974.021

Source: See text.

Table 4-16
Steel Imports Excluded by Tariffs and Nontariff Barriers (Millions of Dollars)

	Estimated Value of Excluded Imports from:										
Importing Country	*Belgium*	*Canada*	*France*	*Germany*	*Italy*	*Japan*	*Luxem-bourg*	*Nether-lands*	*United Kingdom*	*United States*	*Total Imports Excluded*
(2.13)											
Belgium		.354	8 040	12.285	.630	2.007	17.230	3.798	1.267	4.158	59.278
Canada	3.634		3.719	7.529	.653	13.590	1.498	.047	12.644	92.897	168.692
France	24.412	13.855		45.879	2.359	.603	12.273	1.160	2.091	1.058	93.766
Germany	12.395	.123	11.988		4.407	1.570	5.187	4.409	2.780	1.212	104.497
Italy	3.731	.936	8.784	12.704		2.551	1.562	1.447	2.425	2.366	61.610
Japan	.005	.005	.008	.420	.008		.002	.002	.372	13.634	84.900
Luxembourg	4.329	.004	.399	.538	.031	.122		.196	.077	.256	9.689
Netherlands	7.326	.729	2.665	13.536	1.486	.218	3.066		1.117	.287	21.894
United Kingdom	3.861	8.089	6.646	14.918	9.655	6.128	1.616	8.007		16.270	290.146
United States	56.723	155.026	44.428	208.456	39.787	419.636	20.951	11.170	89.077		1593.232

(2.14)											
Belgium		.440	9.781	15.665	.674	2.363	23.179	5.270	1.491	4.262	60.711
Canada	6.607		5.463	11.093	.787	17.191	2.714	.080	15.966	90.297	162.212
France	26.455	.373		47.484	2.347	.613	13.098	1.231	2.127	1.046	93.328
Germany	21.361	.142	17.616		5.527	2.004	8.938	7.337	2.581	1.521	91.225
Italy	5.007	.990	10.415	15.090		2.801	2.095	1.886	2.735	2.324	60.857
Japan	.145	.043	.174	9.197	.044		.058	.091	5.260	27.294	95.619
Luxembourg	6.470	.006	.052	.803	.036	.145		.284	.091	.263	8.959
Netherlands	3.685	.667	1.795	9.099	1.288	.174	1.542		.889	.287	21.412
United Kingdom	14.437	13.162	18.215	41.104	18.420	12.892	6.041	28.063		18.160	314.350
United States	132.274	209.975	101. 897	355.546	68.545	739.103	51.377	31.440	146.503		1870.623
(2.15)											
Belgium		1.392	21.435	33.075	2.858	6.736	34.230	8.034	2.747	99.853	3302.261
Canada	4.303		5.746	11.569	4.772	29.009	1.777	8.718	27.126	1957.506	7863.627
France	35.763	1.384		90.179	25.072	3.062	16.851	4.664	10.695	67.734	2229.670
Germany	16.022	.429	20.445		12.252	3.411	6.705	5.992	6.071	27.020	1326.517
Italy	17.754	12.315	60.848	88.795		17.944	7.429	7.423	17.160	196.766	5737.529
Japan	.005	.006	.009	.454	.010		.002	.003	.412	26.933	584.725
Luxembourg	.113	.021	.773	1.173	.098	.306		.294	.193	7.054	1143.025
Netherlands	18.766	1.578	15.375	75.774	4.809	1.570	7.853		9.274	38.654	3529.690
United Kingdom	4.659	19.109	9.274	20.755	18.209	10.426	1.950	9.906		190.810	5553.844
United States	56.113	239.570	49.491	220.024	37.343	537.298	20.948	11.857	111.935		4240.376

Source: See text.

following an across-the-board elimination of tariffs, the increase in steel imports of Belgium would equal $272 thousand; France, $1.429 million; Germany, $7.762 million; Italy, $2.935 million; Luxembourg, $104 thousand; and the Netherlands, $333 thousand. Japan's excluded imports would also fall to 25 percent of the original estimate, or $12.755 million.

As for nontariff barriers, we noted in Chapter 2 that they are imposed on steel inputs as well as output. Thus estimates of the change in imports caused by eliminating all nontariff barriers (shown in Table 4-12) overstate the increases in the case where countries also remove restrictions on inputs. We can evaluate the degree to which they are overstated by employing the plausible assumption that input and output restrictions have approximately equal nominal rates.

Given the assumptions that $z_i = t_i = t_j$ for nontariff barrier rates, and that $v_i = 0.10$ as before, then equation (2.1) tells us that all nontariff barriers exclude only one-tenth of the steel imports kept out under the alternative assumptions of $t_j = 0$. Therefore, instead of the United States excluding 1 billion dollars of steel as shown in Table 4-12, it may keep out only $100 million. Likewise, for the other major steel traders, the increase in steel imports following elimination of output barriers must be reduced by 90 percent in order to account for simultaneous changes in barriers facing steel inputs. Thus the effective rate of protection accorded by tariffs may be greater than effective nontariff barrier rates in Canada, Japan, the United Kingdom, and the United States, but not in the original EEC.

The Impact of All Trade Restrictions

Returning to our consideration of changes in steel output restrictions only, we can see the net effect of eliminating tariffs and nontariff barriers in Table 4-16. Here it becomes readily apparent that the country whose steel trade balance with the others benefits the most from restricted trade is none other than the United States. American restrictions keep out about 1 billion dollars worth of steel (f.o.b.) from the other major steel traders while the others reduce United States exports by no more than $142 million (c.i.f.).

Certainly, it does not appear likely that differences in trade restrictions caused the American industry's competitive advantage in major world markets to deteriorate. Even before the "voluntary" quota went into effect in 1969, the American steel trade deficit would have been larger by over one-half billion dollars were it not for policies hindering the international exchange of steel. Of course, this partial equilibrium estimate does not imply a similar improvement in the aggregate United States balance of trade. American steel users, paying the higher steel prices that accompany protection, have probably found the competitive position of their exports weakened.

The estimates presented here also take no account of dynamic improvements that might occur in the domestic industry's structure, conduct, and performance under free trade. In the long run, even the steel trade balance might be improved if free trade forced American producers to take the lead in aggressive marketing and technological innovation. Hopefully, these results have demonstrated that consideration of nontariff barriers as well as tariffs is a worthwhile endeavor for policymakers and have suggested new courses of action for them to follow.

Notes

1. W. Kelly Jr., "Nontariff Barriers," *Studies in Trade Liberalization* (Baltimore: The John Hopkins Press, 1967), p. 265.

2. Ibid.

3. The EEC proposes to remove nontariff barriers to intracommunity trade, but progress is very slow. See "Zero Tariff Levels Do Not Make a Customs Union: Commission Proposes Harmonization of Customs Laws and Procedures," *European Community*, no. 109, (February 1968), p. 15; "Jumping the Non-Tariff Hurdle" *European Community*, no. 156, (May, 1972), p. 11. Benelux countries agreed to a deadline of January 1, 1971 for removal of all nontariff barriers, but the extent to which they accomplished this has not been widely reported.

4. Senate Finance Committee, *Steel Imports*.

5. Ibid.

6. Ibid.

7. This estimate for aggregate French import demand is from Houthakker and Magee, "Income and Price Elasticities in World Trade."

8. United States Department of Commerce, *Survey of Current Business* (Washington, D. C.: United States Government Printing Office, February 1970), p. S-1.

9. We present no estimates of the impact of eliminating domestic subsidies on the subsidizing countries exports.

10. These estimates do not reflect the increase in the subsidizing countries' exports due to the subsidy.

11. Tariffs obtained from United States Department of Commerce, *United States Commodity Exports and Imports as Related to Output*, Washington D.C.: United States Government Printing Office, 1963.

12. The share of each material in total inputs is obtained from the United States Department of Commerce *Census of Manufactures*, Washington, D. C.: United States Government Printing Office, 1963.

Part II
Nontariff Barriers to Trade

5 Foreign Trade Policies

The comprehensive list of nontariff barriers developed in Chapter 1 acts as a guide to the barriers which apply to steel imports. The first category of nontariff barriers appearing in this list refers to foreign trade policies: actions taken by governments to influence imports and exports directly. Of these, state monopolization of trade and shipping and insurance regulations appear to be the only policies not applicable to steel. The other eight types of barriers erected by foreign trade policies do restrain the entry of steel into the major steel-trading countries. We consider them below.

Import Licensing

Import licensing systems require importers to obtain permits before purchasing foreign goods. The systems often implement other quantitative restrictions such as quotas, but may also erect important barriers by themselves. In addition, administrative and legal costs of obtaining a license have an effect similar to tariffs.

Working alone and without fixed quotas, import licensing systems may halt imports at any level or time, a fact which increases the risk of importing. Thus a system that stops issuing permits after an arbitrary or unannounced level of imports is reached may be more restrictive than a predetermined quota of the same amount. According to a congressional study, "the mere existence of the licensing requirement provides little clue as to how restrictive the requirement is at a given time, if at all. Indeed, licensing procedures are sometimes maintained only as a 'standby' procedure, in the event that a need should arise for restricting imports."[1]

Whether used alone or in conjunction with quotas, import licenses also allocate the right to import among alternative traders. Issuing one license on an import with no domestic substitutes bestows monopoly power on the licensee, and as a result the volume of imports may be smaller still. This proposition is illustrated in Figure 5-1 where *SS* represents the supply of imports (and with the exception of distribution expenses, the marginal cost schedule for imports), and *DD* the demand for imports. Imposition of quota *OB* causes the import supply schedule to become *SS'*. If the restricted imports are distributed competitively, the quantity of imports will fall by

BC and the import price will rise from *OE* to *OD*. Consider, on the other hand, a license granted to only *one* domestic seller of imported goods with *OB* once again established as the limit on that import. If the import has no close domestic substitutes, a monopoly is established with the volume of imports dropping to *OA* and import price rising from *OE* to *OF*. As long as the quota itself is not too restrictive (i. e., as long as *OB* is larger than *OA*) the monopolization of the right to import will further restrict the volume of imports.

Altogether, nine of the ten main steel-trading countries require licenses for a wide variety of imports, including certain agricultural goods, energy sources, transportation machinery, textiles, chemicals, and metals —excluding steel.[2] Not one of the nine, however, licenses more than a few types of imports. In contrast, Japan subjects all imports, including steel, to licensing in a three-tiered system where permits can be issued automatically, with varying limitations, or according to announced quotas. Although 90 percent of its trade took place under the two less-restrictive types of licenses in 1962, Japan retained quota licensing on 123 items, half of which were manufactures and included passenger cars, computers, typewriters, and power generators. Some alloy steels are subject to quotas which vary from year to year. Moreover, all other steel products are under the second most restrictive type of licensing where unannounced restrictions may be placed on the granting of import permits. According to United States exporters, steel imports to Japan have beeen restricted mostly by this licensing system, although the Japanese have been steadily decreasing the number of items subject to quotas.[3]

Foreign Exchange Restrictions

Import licenses go hand in hand with the exchange permits which importers must obtain before they may make payment for imports. Licenses for foreign exchange could substitute for import permits, since importers cannot purchase goods from abroad without access to foreign currencies. The administrative cost involved in applying for licenses also applies a tariff-like restriction to imports. However, in practice, the right to import and the right to exchange are simultaneously granted.

In the past, many countries have turned to exchange controls to regulate their international payments and/or control imports of "nonessential" goods, but today only developing countries rely on them to a large extent.[4] Nevertheless, three of the main steel-trading countries require exchange permits for all imports, including steel.

Following the crisis of 1968, France imposed exchange controls requiring all import and export payment operations to be conducted directly

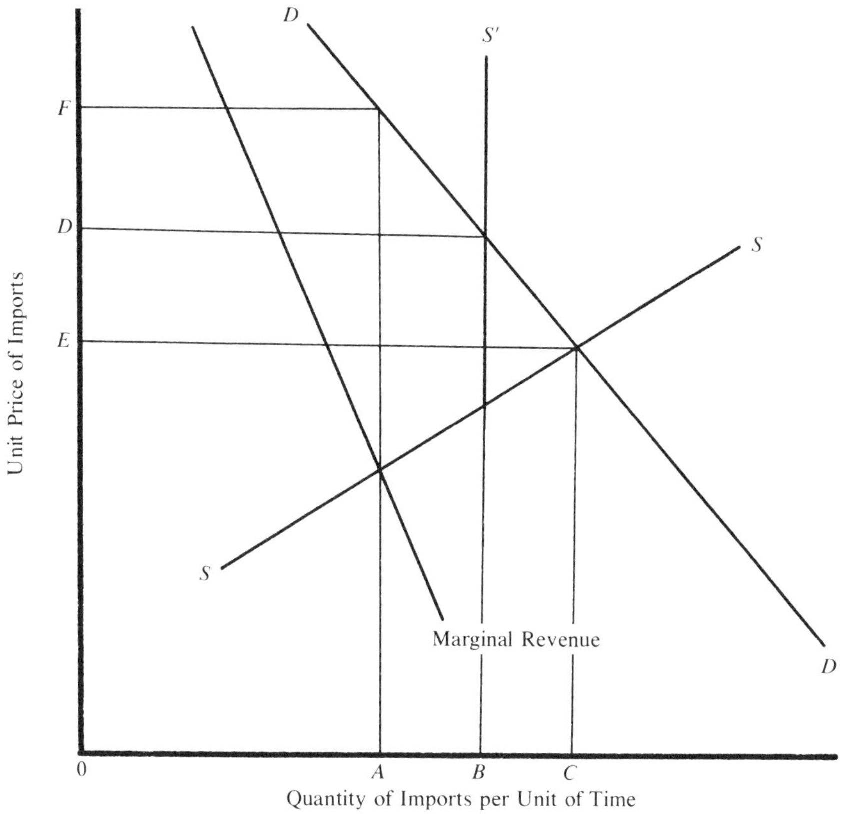

Figure 5-1 The Monopoly Effect of an Import License

through banks authorized to act as intermediaries in foreign transactions. Although certain payments could be made freely through such banks on the basis of general authorization, other transactions faced uncertain treatment.[5] United States steel producers rank exchange restrictions as the fifth most restrictive nontariff barrier to steel in France.

Although Japan generally grants exchange concurrently with import licenses, in certain cases importers must apply for exchange allocation certificates, or obtain approval of their method of payment from the Japanese Ministry for International Trade and Industry.[6] Such exceptions may surreptitiously protect domestic producers. United States steel exporters believe exchange restrictions rank sixth among the nine nontariff barriers they find in Japan. The effect of Japanese exchange controls on United States steel exports has been publicized by William Jackson of the Investors League, who has stated that "it is very difficult somehow to collect Japanese yen for steel sold in Japan.[7]

Although the United Kingdom supposedly issues licenses automatically, there is in fact much room for discrimination in their complex procedure for obtaining foreign exchange.[8] Copies of settlement invoices, exchange control documents, customs entry forms, customs declarations, and the duty charge docket must be submitted to authorized banks when the goods enter at customs. Prior to entrance at customs, the authorized banks will approve payment only upon the submission of import licenses, pro forma invoices, contracts, confirmations of orders, and Sterling transfer forms or import exchange forms (obtained by applications to the government), and documentation which proves that the goods will be dispatched upon payment. United States steel exporters have ranked exchange controls as the prime nontariff barrier to steel in the United Kingdom.

Quotas

Although they also raise the cost of importing in much the same way as tariffs, foreign exchange and import licensing requirements are usually intended to implement quotas. Quotas limit either the value or the quantity of a designated commodity that may be imported into a country within a given period of time. When based on quantity they often result in greater imports of higher-priced items, with the result that the value of imports may not fall in the same proportion as quantity. We also distinguish absolute quotas from tariff quotas which do not permit imports in excess of a certain amount unless an additional duty is levied.

The number of current quotas in the industrial countries is smaller than in the years of currency and wartime disruption, as well as smaller than the number imposed in the less-developed countries. Neverthelesss, all of the major steel-trading countries have quotas on a few products, usually those subject to licensing. A few European countries (France, Italy, and Benelux members) applied quotas to steel in the 1960s, but the quotas were only imposed on selected products (ferroalloys, bolts and nuts) from certain countries (usually Japan). In recent years, only one country has imposed a general quota on all steel regardless of origin.

That quota, which limits French steel imports to 59,000 long tons per month, was imposed by the French after their May, 1968 crisis. Like the other French emergency measures, the quota applied for only a temporary period; July 1 to December 31,1968. The limit was set at an import rate 7 percent higher than that of monthly imports during 1967.[9]

Export Limitations

Akin to quotas are so-called "voluntary" export limitations, negotiated under the importing country's threat to apply more stringent quantitative restrictions. In 1964, there were 67 items subject to "voluntary" quotas in the United States, 28 in Canada, 65 in the United Kingdom, five in West Germany, and one in France, most of which were imposed on Japanese exports to these countries. The United States was the first to use such restrictions when it imposed them on Japanese textile exports to the United States in the mid-1950s.[10]

It also led the way by being the first country to institute "voluntary" restraints on exports of other industrial countries of the West. In January 1969, the United States announced that steel producers in Japan, the EEC, the United Kingdom, and Canada had agreed to impose their own restrictions on steel exports to the United States for the next three years.[11] The agreement came in letters of intent from the Japanese Iron and Steel Exporter's Association and the European Coal and Steel Community, and in less formal assurances from Britain and Canada. Under terms of the agreement steel exports to the United States were limited to 14 million short tons in 1969 and were allowed to expand at a rate of 5 percent in each succeeding year. The limits divided United States steel imports on a 40-40-10 percentage basis among the European Coal and Steel Community, the Japanese, and the two other major steel traders, Britain and Canada, respectively. These allotments corresponded roughly to the percentage shares of these countries in 1967 United States steel imports.

This agreement expired in January 1972, but it was soon replaced by a new, more stringent quota, allowing only 14.5 million tons from Japan, the EEC, and the United Kingdom to enter the United States in 1972. The agreement negotiated between the United States State Department and foreign steel groups also placed strict limits on the tonnage of specialty steels permitted, since foreign producers circumvented the last quota by shipping relatively more of the higher priced products. The pact limits the growth of exports to the United States from signatory countries to no more than 2.5 percent per year, although imports from nonsignatory countries (equal to 2.7 million tons in 1971) can overcome the barriers presented by the "voluntary" export limitations.

United States steel importers rank the new "voluntary" quotas as the most restrictive nontariff barrier erected by the United States. One has estimated that the original quota will result in the importation of higher-valued steel products, with prices averaging 10 to 30 percent higher than those on present imports.

Domestic Biased Procurement

While negotiated export limitations have a quota-like effect on imports, the purchasing practices of governments may possess the restrictiveness of a quota, a tariff, or both. Government procurement policies may, in effect, prohibit or limit imports; alternatively, they may accord a preferential margin to domestic producers on the basis of some percentage of price. Generally, the countries give the same preference to all goods, except where the amount of preference is at the discretion of various government agencies.

Seven of the ten major steel-trading countries, including the United States and six EEC members, provide for methods of awarding government contracts which range from open or public bidding to privately negotiated transactions.[12] In practice, these countries, as well as Canada, Japan, and the United Kingdom often restrict the bidding to domestic firms or enter into negotiations only with domestic firms. Belgium, Germany, Japan, Luxembourg, Netherlands, and the United States specify that domestic goods are to be favored over foreign products in their laws, but Canada, France, Italy, and the United Kingdom have no existing acts imposing general restrictions on federal government procurement from other countries. In the latter countries, however, broad administrative discretion in government purchasing practices affords ample opportunity for discrimination against foreign goods.

Many governments exhibit their bias against purchasing abroad by requiring ministerial approval for large contracts with foreign firms, inviting tenders from national suppliers only, requiring a minimum amount of tenders to come from domestic suppliers, setting aside a certain percentage of government contracts for small domestic businesses, preventing purchasing agencies from contacting firms not legally established in the home country, advising government purchasers "to give special consideration to the value of domestic products," requiring suppliers to maintain a domestic domicile, specifying the domestic origin of certain materials, and entering into exclusive supply agreements with domestic cartels.[13]

It is not surprising, therefore, that world steel traders characterize government purchasing practices as one of the most restrictive nontariff barriers to steel. We do not know the degree of governmental bias in favor of domestic goods, however, except in the five major steel trading countries which accord a preferential margin to domestic producers on the basis of some percentage of price. Three of the five are members of the Benelux Customs Union. Here, a domestic bias exists to the extent that products of foreign origin shall not be used if producers in the Benelux Customs Union can supply the same quality at substantially the same prices. Domestic prices are considered to be substantially the same even if they exceed the c.i.f. import price including duty by as much as 10 percent.[14]

The Canadian government's margin of preference for all domestic goods can be indicated by the percentage preference accorded domestic defense equipment, which is by far the largest single class of public expenditure at the federal level.[15] The price differential on defense spending usually amounts to 10 percent. However, the country follows no hard and fast rule regarding the amount of such preference and no legislation or guidance of a formal character deals with this matter. For nondefense purchases, the premium lies below 10 percent and is based on the difference in price of the foreign content in the tenders. The existing arrangements do not exclude the granting of a preference of more than 10 percent under special circumstances. For coal a premium of up to 20 percent can be paid. Provincial governments give a price differential to domestic goods ranging from 5 to 15 percent. Ten percent is probably the best average since most provinces use this preferential margin including Ontario—the largest province in terms of commerce and population.

The United States also makes an exception to the principle that government departments must award contracts to the lowest bidder. The so-called buy American policy requires that goods of domestic origin (i. e., whose domestic components account for 50 percent of total cost) be purchased by federal agencies for use in the United States except when domestic cost is unreasonable, domestic materials are not available in sufficient quantity or satisfactory quality, or domestic procurement is inconsistent with the public interest.[16] The most important procedure prescribed for the implementation of the Buy American Act determines the standard of reasonableness by which the government evaluates domestic prices. Given the equality of other factors, the United States prices must exceed forcign c.i.f. prices (including duties) by more than 6 percent, or a total of 12 percent applies if the domestic product is produced in an area of substantial unemployment or by a small business firm.

Since 1962 the Defense Department has increased the 6 percent margin of preference to a "benchmark" of 50 percent. However, in practice this price differential averages 28 percent. In no case has the Pentagon applied a preference of more than 50 percent.

As of 1962 the federal government has employed a 50 percent preference for all federal procurement (with the exception of AID expenditures) for use outside of the United States. As for the Agency for International Development(AID), its' predecessors shifted away from a free-world-wide procurement policy to more restrictive policies favoring United States suppliers in early 1960, though AID established no specific price preferences.

Apart from the federal government lies the vast, scarcely charted area of buy american laws, regulations, and undeclared policies in the States and local government.[17] However, most state laws grant preferences to domestic firms by absolutely prohibiting the purchase of foreign steel, rather than by establishing some price differential.

United States steel importers regard domestic-biased government procurement as one of the most important nontariff import barriers in the

Table 5-1
Price Preference Given Domestic Goods by Purchasing Governments[a]

Belgium	10
Canada	10
Luxembourg	10
Netherlands	10
United States Department of Defense	28
Other federal[b]	6

[a]As a percentage of c.i.f. import price.
[b]Excluding AID.
Sources: See text.

country, second only to voluntary export limitations in restrictiveness. We present a summary of the percentage price preferences applied to government procurement in Table 5-1.

Subsidies

In general, any subsidy restricts imports by giving domestic producers a competitive advantage vis à vis foreign suppliers. We classify subsidies as direct or indirect, according to whether they take the form of a direct cash payment on the one hand, or the guise of tax exemptions, low-cost loans and other services, reduced raw material and labor costs, or monpolistic privileges which permit higher output prices on the other.[18] Alternatively, we classify bounties as domestic or export subsidies, depending on whether government aid goes to all domestic production or just to exports. Although domestic subsidies sometimes have the purpose and always the effect of influencing the foreign trade sector of an economy, we consider them in Chapter 7 on internal economic policies.

Among the major steel trading countries, indirect subsidies have been favored over direct ones, although steel in particular does receive some cash payments. The subsidies include accelerated depreciation, investment incentives, low interest rates, "over-priced" government purchases, internal tax rebates, subsidies to factors of production which lower their prices, direct subsidies to exporters, overseas marketing assistance, tied foreign aid, and border tax rebates. Only the last four are pure export subsidies; the others will be considered later. All these subsidies apply to steel, as well as other goods.

Direct Export Subsidies

France, the only major steel-trading country to make cash payments to exporters, started granting indemnities to exporting manufacturers of industrial products in order to compensate for increases in their wage costs following the 1968 crisis. The original "basic rate of compensation" was set at 6 percent of export production costs. France's European neighbors also appear to subsidize shipment of exports on domestic transportation facilities over which they have control, although we do not know the exact amount of subsidization.[19]

The Japanese government grants a percentage increase in depreciation allowances and defers the direct profits taxes of exporters.[20] The benefit derived from such deferrals equals the alternative cost of obtaining capital saved by using deferred taxes as a substitute source of funds, and this constitutes a pure export subsidy.

Overseas Marketing Assistance

While the GATT prohibits direct export subsidies, most countries apply various indirect measures to assist in the marketing of all exports.[21] The major selling expenses faced by exporters include insurance, credit, product promotion, market information, and penalties for failure to export.

Credit. Credit granted by exporters in each of the major steel-trading countries can be insured at a nominal premium with their respective governments.[22] Similarly, all ten countries will directly finance exports or refinance credit granted by their exporters. In the case of Britain, Belgium, France, Italy, and Japan, exporters may discount their receivables at a preferential rate, although the first two countries do not disclose the precise amount. For the rest, however, the difference between the preferential and market discount rate can be used to quantify this export subsidy.

French exporters, for instance, may discount their receivables with the Bank of France at a rate of 5.8 percent. This contrasts with the regular discount rate of 8.5 percent.

Mediocredito is an agency of the Italian government which will refinance up to 85 percent of export credit at an interest rate of 3 percent. The balance of the credit is refinanced at the general interest rate of 8.5 percent. Thus Italian exporters receive an ad valorem export subsidy of 4.7 percent.

The Export-Import Bank of Japan finances 80 percent of long-term export credit on steel and other capital goods at an interest rate of 4 percent, considerably lower than the 5.5 percent charged for credit on other manufactured goods. Thus the Japanese export subsidy equals 1.2 percent.

Product Promotion and Market Information. All ten major steel traders engage in product promotion, consisting of advertising exports; designing, erecting, and financing trade fairs and exhibits; sponsoring trade missions; and covering losses arising from expansion into new foreign markets. Likewise, every country conducts market research and provides export marketing information to domestic firms. In addition, Canada has a credit reference service for exporters, Japan offers her exporters free consulting engineering services, and France sponsors trade conferences.[23]

Penalties as Export Incentives. The Japanese government not only offers overseas marketing assistance, but also offers the most direct incentive to export of any major steel trader. Since GATT forbids direct export subsidies, Japan has offered its steel industry an incentive to export by penalizing those firms which do not. The Japanese government is able to impose penalties through its close control of Japanese cartels in which such inputs as investment capital and raw materials are allocated among firms, and production or sales quotas are established and enforced by fines.[24]

At the end of 1967, the Ministry for International Trade and Industry (MITI) informed members of the Japanese steel industry that failure to export 10 percent more tonnage in 1968 would result in a production penalty of $28.00 per ton and curtailment of their coke and coal supply. Such a sizable penalty ($28.00 per ton of production equals $140.00 per ton of exports in 1966) would have caused Japanese supply (*SS*) in Figure 5-2 to kink at the price where Japanese exporters would normally be willing to supply 110 percent of 1967 export tonnage. Below that price the new supply curve (*SS'*) would be perfectly inelastic, while above that price the function would retain its original elasticity. The volume of exports would rise and export prices would fall in much the same way they would behave with a subsidy.

Tied Foreign Aid

The tying of economic development grants to purchases in the donor country may oblige the recipient to pay higher prices than it could obtain on the world market. Thus tied foreign aid may bestow a monopolistic privilege, and in effect a subsidy upon producers in the donor country. Aid commitments as a percentage of gross national product are largest in the United States and France. Moreover, these countries are also foremost in the restrictive tying of aid, though the resulting subsidization has not been quantified.[25]

Border Tax Rebates

Introduction. Border tax adjustments generally purport to place imports

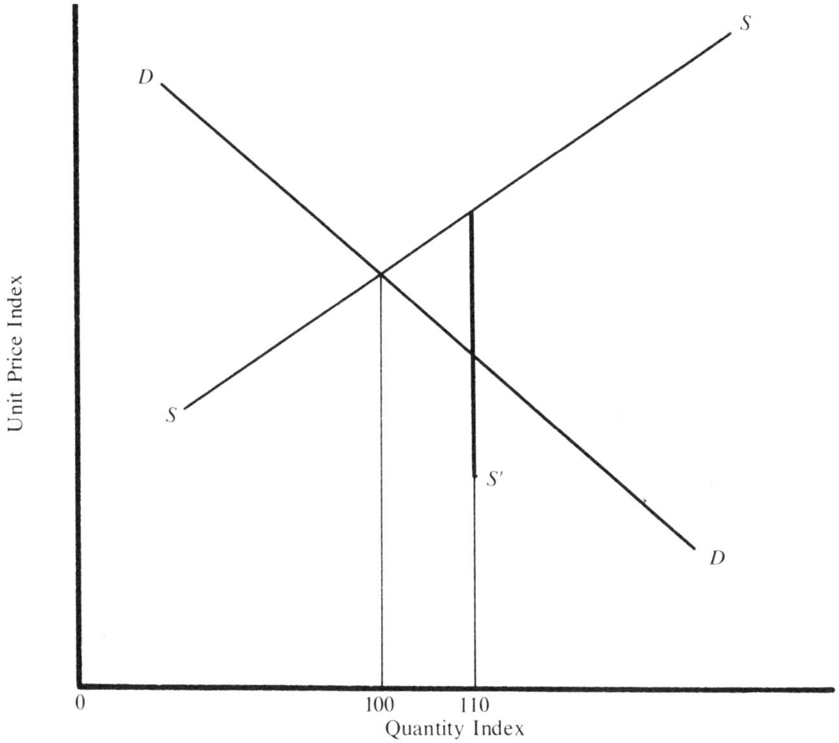

Figure 5-2 The Japanese Steel Export Market (1966 Export Volume = 100)

on the same footing as domestically produced goods with regard to taxes imposed on them. The GATT has legitimatized the border-tax practices of Western countries by sanctioning adjustments for indirect taxes in excess of their nominal rate.[26]

No widespread agreement exists over the economic distinction between direct and indirect taxes, but we can safely ignore conceptual problems of definition.[27] The real world leaves the job of categorizing taxes to GATT officials. They have decided that any levy on a particular good "at one or several stages" of production and distribution is indirect, and that net corporate income taxes are direct.[28]

Indirect or sales taxes may be classified, according to the manner in which they are collected, as (1) single-stage, (2) value-added, or (3) cascade taxes. Single-stage taxes are collected at only one point in the chain of distribution, while cascade taxes are assessed on the value of each transfer of goods. The value-added type effectively taxes only the increase in the worth of product over the cost of material inputs, but in practice the levy applies to the full value of a sale; the taxes on the purchase of inputs (tax occulte) are then deducted.[29]

The Direct-Indirect Tax Mix. The American iron and steel industry, among others, contends that "the United States is more dependent on direct taxes than on indirect taxes."[30] Moreover, it claims that under current GATT rules "a country with a high ratio of direct to indirect taxes is at a disadvantage in export markets when competing with a country whose fiscal system is weighted in favor of indirect taxes."[31]

Evaluation of the first statement that the United States relies more upon direct taxation presents difficulties because of the different bases to which taxes can be related. It is true that using the total tax bill as a base (see Table 5-2) shows a stronger dependence on direct taxes in the United States than in any other major steel-trading country, save Japan.[32] Using GNP as a base (as in Table 5-3), the United States still appears to rely relatively more on the nonadjustable direct taxes than most countries, but the differences become less pronounced. However, in order to compare the actual tax burden of steel companies in the ten countries considered here, net corporate income may be a more appropriate measure. By this base, the direct tax burden on United States corporation almost equals that imposed on most foreign companies, as shown in Table 5-4.

The Effect of the Tax Mix on Trade. Given that the United States may rely slightly more on direct taxes than on indirect taxation, the question remains: Does this nonrebatable burden inhibit United States exports and encourage United States imports? Those who answer affirmatively argue as follows: If direct taxes are shifted forward, they raise the price of the good to the buyer. If the United States depends more on direct taxation, but can only exempt exports from indirect taxes, its export prices will be higher than those of Europe. Europeans who depend more on indirect taxes can rebate more on their exports. Moreover, United States goods face higher indirect taxes at the borders of foreign countries than foreign goods face upon entrance to the United States.

Three tenuous assumptions underly this line of reasoning: (1) The incidence of taxes is the same in each country; (2) the incidence of taxes is the same for imports, exports, and domestically traded goods; and (3) the tax burden shifted forward in the case of direct taxes equals or more than equals the tax burden passed forward in the case of indirect taxes. The first assumption requires that demand and supply elasticities for all goods relate to each other in the same way in each country. The second assumption depends on the foreign trade demand and supply elasticities for each country satisfying the same exacting relationship as the domestic elasticities.

The third assumption must hold in order for a direct tax rebate to lower export prices in the same way that an indirect tax rebate does. Empirical tests of this proposition produce differing results, none of which indicate

Table 5-2
Direct and Indirect Taxes as a Percentage of Total Taxation[a]

	Profit Taxes	*Consumption Taxes*
Belgium	6.5	36.9
Canada	16.0	33.1
France	4.8	34.1
Germany	11.2	28.8
Italy	2.0	33.0
Japan	23.2	27.9
Netherlands	8.1	24.8
United Kingdom	6.7	34.7
United States	16.0	19.6
Luxembourg	NA	NA

[a]1963-1965 average.

Source: OECD, *Border Tax Adjustments and Tax Structures in OECD Member Countries*, Paris: 1968, p. 198.

Table 5-3
Direct and Indirect Taxes as a Percentage of GNP

	Profit Taxes	*Consumption Taxes*
Belgium	1.8	10.2
Canada	4.2	8.7
France	1.8	12.8
Germany	3.8	9.7
Italy	6.0	10.2
Japan	4.5	5.4
Luxembourg	NA	NA
Netherlands	NA	8.3
United Kingdom	2.0	10.3
United States	3.9	4.8

Source: OECD, *Border Tax Adjustments*, p. 199.

Table 5-4
Profits Tax as a Percentage of Corporate Profits

Belgium	36
Canada	50
France	50
Germany	66
Italy	48
Japan	35
Luxembourg	40
Netherlands	47
United Kingdom	40
United States	48

Source: OECD, *Border Tax Adjustments*, p. 195.

the incidence of direct and indirect taxes to be equal, and only one of which finds the incidence to be higher. Zero shifting has been found by M. A. Adelman[33] and C. Hall,[34] while 100 percent shifting has been found by Musgrave and Kryzyzniak.[35] A debate, too lengthy to summarize here, has ensued over the reliability of these empirical tests.[36] As for this study we shall retain the traditional theoretical notion that there is zero shifting in the short run and ignore the doubtful impact of small differences in the direct-indirect tax mix. Balassa concurs with this evaluation when he concludes that American reliance on direct taxes "does not discriminate against American exports to an appreciable extent."[37] This conclusion should not prejudice our results to any significant degree, since even under the assumption of equal incidence for direct and indirect taxes, rebates of direct taxes may not affect relative export prices. Using taxes as a percentage of GNP for 1961, Aliber and Stein found less than a 2 percentage point difference among the direct taxes of France, Germany, Japan, the United Kingdom, and the United States. They found that the rebating of direct taxes on United States exports would lower their price by almost 5 percent, assuming an equal incidence for direct and indirect taxes. However, if other countries also rebated direct taxes, the relative prices of United States exports would decrease by considerably less: 1 to 2 percent.[38]

A more recent contribution, it should be noted, estimates that if we include social security taxes with other nonrebatable direct taxes, the United States possesses a competitive advantage under the existing border adjustment system. On the other hand, the same study also finds that the present European border adjustments for indirect taxes have placed United States exports at a competitive disadvantage in foreign markets. Even though domestic- and border-tax rates may be equal, different elasticities of supply and demand apparently cause distortions in trade flows.[39]

Rebates in Excess of Indirect Taxes. Border-tax adjustments may unambiguously distort trade patterns in two ways. Firstly, when the border tax imposed on imports exceeds that assessed on similar domestic products at the same stage of distribution, there may be a tariff-like restriction of imports. Secondly, when the border-tax rebate granted exports is more than the tax levied on domestic sales of the same good, there may be a pure export subsidy. We consider import border taxes in a later section on internal economic policies since their purpose, and for the most part their effect, is to levy the same internal tax on all goods. Border-tax rebates on exports, on the other hand, intend to at least equate the competitive position of a country's exporters with that of foreign suppliers, and may be appropriately considered in this section on foreign economic policies. Almost all of the cascade-tax countries considered here lack any uniformity between their rebates to exports and extra charges on imports, due to the arbitrary manner of calculating border-tax adjustments.

The GATT's opposition to export subsidies is implicit in its recognition that "a subsidy on the export of any product may have harmful effects for other contracting parties both importing and exporting . . . and may hinder the achievement of the objective of this agreement."[40] The treaty does not consider exemption of taxes levied on similar domestic goods to be a subsidy, nor does it permit tariff retaliation by another country to compensate for refunds of taxes imposed on an exported product. It is these rebates (if excessive) which consititute pure export subsidies, and we begin examining the occurrence of this type of subsidization among the major steel-trading countries by exploring the indirect tax systems in each of the nations under consideration.

Belgium, Italy, Luxembourg, and the Netherlands usually applied cascade taxes in the 1960s, though they planned to switch to value-added taxation systems in 1972 in accordance with EEC tax harmonization. France and Germany apply value-added taxes, while Canada, Japan and the United Kingdom have imposed single-stage taxes at the wholesale or retail level. The United States levies only a few special excise taxes at the federal level, while the states apply both single-stage taxes at the retail level or cascade taxes on every transfer of property.

Usually countries do not levy single-stage taxes on steel exports, but if they do, rebates generally equal the tax and cause no subsidization. Steel exports usually escape value-added taxes also, and only the amount of tax actually paid on inputs is refunded to the exporter, so that once again exporters receive no bounties. Rebates to steel exporters for cascade taxes, on the other hand, usually subsidize exports and create corresponding nontariff barriers to imports. This conclusion follows from detailed consideration of the taxes, rebates, and industrial structure in the four cascade-tax countries mentioned above.

The border-tax rebate of cascade taxes on any good constitutes an export subsidy for two reasons.[41] First, it often includes taxes on purchases of plant, equipment, power, and services while the GATT Article VI provides that only taxes on the product itself may be rebated. Second, the amount of tax remitted is usually determined in accordance with arbitrary tax remission rates established by the cascade-tax countries as distinguished from an exact accounting by the applicant of the amount of taxes in fact paid on the materials used in producing the goods for export. In many cases the amount of tax refunded under this formula exceeds the tax paid by the manufacturer with respect to the raw materials purchased by him for use in manufacturing the exported merchandise. The frequency with which rebates tend to exceed tax occulte in Europe is reported to have created many problems within the Common Market.[42]

For steel in particular, the border-tax rebates result in export subsidization because there are few if any actual payments of tax occulte to be compensated for. The high degree of vertical integration in steel makes taxable purchases of raw materials uncommon. Moreover, the few inputs actually purchased are generally subject to extremely low or nonexistent taxes. Finally, inputs for steel destined for export are usually exempted from all taxes and duties. We can find evidence below supporting each statement in this and the preceding paragraph. Rebate and border-tax rates refered to below are expressed as a percentage of c.i.f., duty-paid value, excluding the tax charges.

Rebates for Taxes on Plant, Equipment, and Power. Article VI of the GATT provides that "no *product* . . . imported into the territory of any other contracting party shall be subject to anti-dumping or countervailing duty by reason of the exemption of *such product* from duties or taxes borne by the *like product* when destined for consumption in the country of origin . . . or by reason of the refund of such duties or taxes."[43] Nevertheless, under the Belgian and Luxembourg cascade-tax systems, rebates included taxes paid on purchases of machines and equipment, estimated property taxes, as well as taxes paid on processing, transportation, and rental expenses concerned with the handling of goods prior to exportation. The rebatable Italian taxes included levies on the furnishing of heat, light, and power, as well as those on sales of the product at various stages of production and distribution. In the Netherlands, provision was made for rebates of taxes paid on solid fuels, electricity, gas, capital equipment, and services in the manufacture of exports.[44]

Arbitrary Rebates. In 1969 Belgian steel exporters, already exempt from the sales tax on their finished product, received a lump-sum export rebate averaging 7 percent. Since the amount of the rebate does not vary directly with the degree of finishing, no clear relation between tax occulte and export rebates seems to exist.[45] Similarly, Dutch rebates, ranging from 0.75 to 10.4 percent, are paid to exporters to compensate for "estimated" tax occulte. Exports from the Netherlands were not only exempt from the standard tax of 5.24 percent, but also received a rebate averaging 5.57 percent. Luxembourg applied an across-the-board rebate of 1 percent to all products, including steel, regardless of their stage of manufacture.[46]

The Italian cascade-tax system subjected iron and steel mill products to a cumulative turnover tax of 4 percent, and Italian steel exporters received a 4 percent refund. To compensate for *estimated* taxes on inputs, however, Italy granted additional export rebates which varied from 3.6 to 7.8 percent for steel, but most commonly amounted to 4.8 percent of the export price. The rates of yet another rebate on iron and steel mill products varied from Lire 15 to Lire 30 per kilogram. This specific rate is the equivalent of $25 to $50 per metric ton, or an ad valorem rate of 18 to 35 percent.[47]

The first Italian compensatory export rebate supposedly equalled an estimated average of taxes paid.[48] Nevertheless, a congressional study found the refund to be excessive.[49] As for the second, specific, export rebate, it is calculated on a weight-basis without regard to the taxes paid.[50] Adding the minimum specific rebate (in ad valorem form) to the first ad valorem rebate, one obtains a total export subsidy of 22.8 percent due to border-tax rebates.

Referring to two of the cascade-tax systems, specifically, the OECD concluded that "it is quite possible that the tax refunded on particular products to certain firms might be higher than the average tax paid on these products on the home market."[51] With respect to steel this appears to be an understatement.

Vertical Integration in Steel. Under the cascade-tax system, the amount of tax paid by a producer may vary greatly from one firm to another, depending on the degree of vertical integration.[52] Since most steel comes from fully integrated manufacturers in the cascade-tax countries,[53] any refund of "estimated" tax occulte seems unjustified. Many of these large steel complexes are export-based, since they sell only a small share of their output domestically. With respect to such firms the OECD concluded that "the export intensive firm, being generally larger and more integrated than the average domestic firm selling similar products is likely to have paid less tax on each product than the average firm on the home market."[54]

Low Input Taxes. The four countries considered here subject many inputs to very low tax rates or exempt them from taxes altogether. Belgium imposes only a 1 percent lump-sum tax on such important raw materials as iron ore and coal,[55] and certain services carry a reduced tax rate of 0.7 percent. Moreover, the OECD reports that any raw materials or semifinished products can be exempted from transmission taxes if they eventually leave Belgium, Italy, and the Netherlands. Italian materials may bear a tax rate as little as 0.6 percent, fuels used in steelmaking as little as 1 percent, and purchases of capital assets are exempt from taxation. Luxembourg levies a 0.75 percent tax on fuel oil, and 3.75 percent tax on coal, but exempts purchases of gas, water, electricity, and some services and raw materials from taxation. In general, the Department of Commerce reports that Netherland taxes raw materials at a rate of 1 percent, while fuels such as coal are exempt from Dutch tax.

Exemptions from Import Duties. All four of the major steel traders which used cascade-tax systems also import much of their steel-making raw materials. To the extent that these raw materials are destined for export in the form of steel, the four governments exempt them from all taxes and duties.[56] Nevertheless, this exemption does not appear to affect the estimated export rebates in these countries. The estimated rebates often include an amount to compensate for duties paid on imported inputs.[57] Italy, for instance, grants a rebate of "*presumed* customs duties and indirect taxes (other than the general turnover tax) paid on the raw materials. . . ."[58]

Conclusion. As the preceding pages point out, arbitrarily determined rebates to steel exports for nonexistent payments of tax occulte seem totally unjustifed. In the case of steel exports under the four cascade-tax systems mentioned above, in fact, the input tax rebates appear to be pure export subsidies. Of course, these countries have been in the process of moving toward value-added tax systems, and the value-added tax countries give refunds only for those input taxes actually paid. Therefore, no pure export subsidy results. On the other hand, one single-stage-tax country, the United Kingdom, does subsidize steel exports with a rebate of about 3 percent. Because the steel rebate has the official purpose of promoting overseas sales, and is granted in addition to the export refunds for specific input taxes, the 3 percent rebate can be looked upon as a pure export subsidy.[59] In summary, Table 5-5 shows the extent to which border rebates subsidized exports from each of the major steel-trading countires.

Import Surcharges

The American steel industry claims the EEC countries dealt with their steel import problem by imposing import surcharges.[60] However, we find no evidence of surcharges in the original Common Market countries, except in the case of Luxembourg, which has imposed a 3 percent surcharge for some time.[61] United States steel producers, themselves, erected a barrier of uncertainty to United States steel imports by publicly asking Congress for a temporary import surcharge on steel in 1967.[62] Their action supplemented the industry's standing request for an import quota. And on August 15,1971, they got their wish, as President Nixon instituted a four-and-one-half-month period of surcharging American imports. Steel became the only quantitatively restricted import subject to the 10 percent surcharge.[63]

Nevertheless, this discussion should not gainsay the fact that many other countries have imposed import surcharges in the past. The most recently retracted levies were in Canada and the United Kingdom. Starting in October 1964, the British imposed a surcharge of 15 percent on all imports except certain foodstuffs, and raw materials, and the levy was not eliminated until November 1966. However, with regard to solving the British balance-of-payments problems, "the surcharge was not very effective under conditions of overfull employment and excess demand."[64]

Table 5-5
Export Subsidization Due to Border Tax Rebates[a]

	Excess Border Tax Rebate
Belgium	7.00
Canada	0.00
France	0.00
Germany	0.00
Italy	22.80
Japan	0.00
Luxembourg	1.00
Netherlands	5.57
United Kingdom	3.00
United States	0.00

[a]As a percentage of export price.
Source: See text.

In June 1962, Canada imposed surcharges of 15, 10, or 5 percent on products affecting about one-half of its total annual imports, but eliminated the levy on March 31, 1963. The 5 percent surcharge applied to most steel products. The Canadian surcharge greatly reduced imports during its brief operation, due to "the slack" in the Canadian economy.[65]

Antidumping Measures

Jacob Viner's classic work on the subject defines *dumping* as "price-discrimination between national markets."[66] Dumping commonly refers to sales of a good to foreign purchasers at a price lower than the price of the same goods to domestic buyers. Although governments appear to be more concerned with the effect of low-priced, dumped imports on domestic producers, the main economic objection to dumping relates to the imperfectly competitive conditions which cause it. In fact, dumping has the economic advantage of giving import buyers the benefit of lower prices. Although the dumping country discriminates against its own buyers, their prices need not be higher, and even if they are, this decrease in the real income of consumers must be weighed against the increased profits of producers. The uncommon occurrence of predatory dumping to force out competitors is the only time when economics does not justify dumping. Still, governmental preoccupation with protecting domestic producers has resulted in the proliferation of regulations that authorize antidumping duties. Apart from countervailing duties, the administrative procedures implementing antidumping regulations may, in themselves, hinder imports through delay and uncertainty.

The GATT recognizes the harm to domestic industry caused by dumping, and it permits contracting parties to levy countervailing tariffs to make up the difference between the dumped-import price and the normal price. The normal price is defined as either the comparable domestic price in the exporting country, or in its absence, the export price of a third country, or the cost of production and selling plus a reasonable profit margin in the exporting country. Due allowance must be taken for legitimate differences in price, and no duty may be levied, except in the case of subsidies, unless material injury is expected to befall an established or infant industry. Since the determination of possible injury usually involves some delay, the importing country may make an immediate, but temporary, countervailing levy.[67]

For the most part, antidumping regulations in the major steel-trading nations conform with the GATT provision.[68] There are a few inconsistencies in government practices and loopholes in the GATT, however, some of which will be rectified or closed by the Kennedy Round Antidumping Code. These problems will be taken up below, together with a consideration of how steel has fared regarding antidumping measures. Balassa considers steel to be "one of the foremost candidates for dumping" because of the "prevalence of oligopolies" in the steel industries of many countries.[69]

As part of European economic integration, the EEC plans a common antidumping policy.[70] Until recently, however, most European countries had no antidumping laws. No specific antidumping regulations exist in the Netherlands, but measures can be taken under the customs tariff procedures set up by the Benelux Treaty. These procedures provide for countervailing duties of 10 percent minimum to twice the regular tariff rate at a maximum on goods from countries which maintain a less favorable policy toward (or whose policy interferes with the vital interests of) the Benelux countries. Similarly, the Benelux agreement guides the antidumping policies of Belgium and Luxembourg, though Belgium has also incorporated the provisions into law. The GATT found the formal legal power in Belgium to be wider in scope than GATT permits, but Belgium claims that it respects GATT restraints on antidumping practices in executive orders. Another restrictive effect of the Belgian system is that importers have no opportunity to present their side of the case. Adoption of the new Kennedy Round Antidumping Code, however, will supposedly ensure fair and open hearing where importers can defend their interests. Out of the ten antidumping complaints lodged by Belgian producers between 1948 and 1958, only three were subjected to antidumping measures and none of the three goods were steel products.

Canadian antidumping provisions are unique in that an undefined, normal duty valuation provides a basis for determining the margin of dumping.[71] The penalty for dumping consists of a countervailing duty which may not exceed 50 percent of the normal market values. We find another unique aspect of Canadian law in the provision that no injury to domestic industry need occur for antidumping duties to be levied. Canada does not report the number of requests for antidumping restrictions nor the times they are granted. Therefore, no public information exists on restriction of steel imports in particular nor of imports in general, except for the Canadian-American Committee's estimate that "Canada has not made extensive use of the valuation process as a deliberate means of restricting trade. . . ."[72] The GATT finds Canadian antidumping provisions narrower in scope than its own, except insofar as the injury requirement is concerned. Now, Canada has in effect agreed to adopt an injury requirement by accepting the Kennedy Round Antidumping Code.[73]

France, Germany, and Italy, all have laws which delegate general powers concerning antidumping measures to their respective customs' authorities.[74] The three countries' customs codes authorize imposition of countervailing duties under the same circumstances provided by the GATT: below-normal prices or subsidies, and and serious injury.[75] Recent modification of French law, however, increases the applicability of antidumping measures by narrowing the definition of an industry in cases claiming serious injury.[76]

Germany has taken no formal action against steel, but German officials forced one United States steel company to recently raise its prices with the threat of antidumping duties.[77]

The Japanese possess no specific antidumping legistation, though the Customs Tariff Law of 1910 offers this type of protection. Article 9 of the law provides relief along the same lines as GATT but allows the government to impose countervailing duties retroactively—a practice which increases the uncertainty of importing.[78] We find no information available on the extent to which the Japanese use antidumping measures to stop steel imports.

The British levy countervailing duties only after the House of Commons confirms a Board of Trade order.[79] Usually the Board of Trade awaits complaints from domestic industry before investigating or initiating action against dumping, according to GATT provision. Such a complaint (concerning steel imports) has been publicly contemplated by the British Iron and Steel Federation recently. Between 1961 and 1964, twenty-nine antidumping complaints were made by various British industries, but none of these formal complaints involved steel.

While the United States antidumping legislation generally conforms with GATT rules, importers complain about delays and administrative practices.[80] The GATT has found United States legislation to be wider in scope than its own regarding the injury requirement, but Europeans report that in practice less stringent injury requirements have been applied only to agricultural products. United States law also empowers the administration to levy countervailing duties on subsidies granted by private parties such as cartels, while the GATT restricts them to goods subsidized by governments, except for cases of material injury. Neither of these inconsistencies were apparently corrected by the Kennedy Round Antidumping Code, whose main impact on the United States has been to require injury adjudication and thus the withholding of appraisement to last no longer than ninety days.

Table 5-6 summarizes information on the overall number and disposition of United States antidumping cases. Although the data overlap, they demonstrate that antidumping measures do not restrict imports primarily through imposition of countervailing duties. The government withheld appraisement in almost one-half of the most recent cases and importers did not know what profit if any to expect on their sales, even though in most cases antidumping duties were never assessed. In approximately one-fourth of the cases, the threat of antidumping restrictions has forced import price increases. Between 1948 and 1957, one of two antidumping actions related to cast iron pipe from the United Kingdom.[81] According to the Canadian-American Committee, the action was based on injury to "one marginal producer" who "had not been making a profit even when there had been no imports." By the committee's reckoning, imports equalled less than "four-tenths of one percent of domestic production."[82]

During the period 1958-64, steel imports accounted for one-fourth of the antidumping restrictions applied. Two 1964 actions imposed antidumping duties on steel reinforcing bars from Canada, and on carbon steel bars and shapes from the same country.

Two years before these events, however, American producers entered a more publicized dumping complaint against imports of steel wire rods from the EEC and Japan. Although imports from the EEC were found to be sold at "lower than fair value," the Tariff Commission found that no American industry was being injured as a result of the dumping.[83] Nevertheless, the number of complaints, and the uncertainty and delay accompanying them, caused United States steel importers to rank antidumping measures in third place among America's nontariff restrictions.

To illustrate the high rate of protectionist activity in this area, 1973 witnessed three new cases. In January of that year, the Anti-Friction Bearing Manufacturers Association requested "tariff adjustments" for its foreign competition, and Latrobe Steel filed an antidumping complaint against Swedish steel producers. In February 1973, the Customs Bureau ruled that Sweden sold stainless plate at less than "fair value" in an antidumping complaint filed by Jessop Steel Company.[84]

Table 5-6
United States Antidumping Cases

Calendar Years	*Number of Cases*	*No Price Discrimination*	*Price Revision*	*No Injury*	*Injury*	*Appraisement Withheld*
1948-57	96	NA	NA	NA	2	NA
1955-63	282	187	62	27	6	NA
1959-64	194	108	53	25	8	89

Sources: 1st row—GATT, *Basic Instruments and Selected Documents*, p. 117.
2nd row—Hemmendinger, *Non-Tariff Trade Barriers*.
3rd row—*United States Congressional Record*, June 1, 1965.

Notes

1. Subcommittee on Foreign Economic Policy, Joint Economic Committee, 87th Cong., 1st sess., *Trade Restraints in the Western Community* (Washington, D. C.: United States Government Printing Office, 1961), p. 11.

2. Data on licenses comes from Joint Economic Committee, 89th Cong., Hearings before the Subcommittee on Foreign Economic Policy: *The Future of United States Foreign Trade Policy* (Washington, D. C.: United States Government Printing Office, July 1967), pp. 311, 314, 317, 321, 331, 332, 335, 341; *Foreign Trade and Tariff Proposals* House Ways and Means Committee, 90th Cong., 2nd sess., (Washington, D. C.: United States Government Printing Office, 1967), pt. 9, pp. 3944, 3901, 3902, 3903; F. Masson and J. Whitely, *Barriers to Trade between the United States and Canada*, Canadian-American Committee, 1960, p. 22; United States Treasury Department, Bureau of Customs, *United States Imports Requirements* (Washington, D. C.: United States Government Printing Office, 1965), p. 6.

3. United States Department of Commerce, ''Japan Lifts Import Restrictions on 39 Additional Commodities,'' *International Commerce* (Washington, D. C.: United States Government Printing Office, October 26, 1964), p. 14.

4. International Monetary Fund, *19th Annual Report on Exchange Restrictions*, Washington, D. C., 1968.

5. United States Department of Commerce, ''France Reimposes Exchange Controls in Wake of Crisis,'' *International Commerce*, July 7, 1968, p. 22.

6. *Future of United States Foreign Trade Policy*, p. 308; and *Overseas Business Reports*, 65-11.

7. House Ways and Means Committee, *Foreign Trade and Tariff Proposals*, pt. 7, p. 1034.

8. Joint Economic Committee, *Future of United States Foreign Trade Policy*, p. 310; and *Overseas Business Reports*, 68-83, p. 6.

9. "The Community Deals with French Crisis," *European Community* (August 1968), p. 3.

10. Kelly, "Nontariff Barriers," p. 210; and K. Bauge, "Voluntary Export Restrictions as a Foreign Commercial Policy with Special References to Japanese Cotton Textiles," Ph. D. thesis, Michigan State University, 1967.

11. A complete description of the 1968 and 1972 agreements appears in W. T. Hogan, *The 1970s: Critical Years for Steel*, (Lexington, Mass.: D. C. Heath, 1972), pp. 45-72.

12. We obtain information in this paragraph from *Congressional Record*, (Washington, D. C.: United States Government Printing Office, April 28, 29, May 9, 1965), pp. 8716-21; "Cabinet Order No. 336," *Official Gazette of the Japanese Government*, September 25, 1963., *Translation for Congressional Record*, May 9 1965, p. 9706; Federal Property and Administrative Services Act of 1949, 41 *United States Code* 251 et seq.; Armed Services Procurement Act of 1947 (now Chapter 137 or Title 10 of the *United States Code*.); House Ways and Means Committee *Foreign Trade and Tariff Proposals*, pt. 1, p. 225; OECD, *Government Purchasing in Europe, North American, and Japan: Regulations and Procedures,* Paris: 1966, pp. 38, 56, 65, 103-115; M. Massel, "Nontariff Barriers"," pp. 65, 67, 68; N. Hemmindinger, *Non-Tariff Trade Barriers of the United States* (Washington, D. C.: United States-Japan Trade Council, 1964), pp. 8-12; M. Massel, *Competition and Monopoly; Legal and Economic Issues*, Washington, D. C., The Brookings Institution, 1962, p. 64; Kelly, "Nontariff Barriers," pp. 278-79; L. Towle, International Trade, pp. 544-47; Subcommittee on Defense Procurement, Joint Economic Committee, 88th Cong., 1st sess., *Hearings on the Impact of Military Supply and Service Activities on the Economy,* Washington, D. C.: United States Government Printing Office, 1963; H. van Cleve, "The Use of Federal Procurement to Achieve National Goals," *Wisconsin Law Review*, (1961), pp. 566, 577-92; L. Knapp, "Buy-American Act: A Review and Assessment," *Columbia Law Review* (1961), p. 430; P. Gantt and W. Speck, "Domestic v. Foreign Trade Problems in Federal Government Contracting: Buy-American Act and Executive Order," *Journal of Public Law* (1968), p. 378; Masson and Whitely, *Barriers to Trade between Canada and the United States,* pp. 39-45.

13. This paragraph is based on B. Balassa, *Trade Liberalization among the Industrial Countries* (New York: McGraw-Hill, 1967), p. 65: "Euro-

pean Non-Tariff Obstacles to Trade Criticized," *European Community* (August 1965), p. 8; OECD, *Government Purchasing* pp. 115, 36, 38, 61, 72; *Congressional Record*, April 29, 1965, p. 8903; House Ways and Means Committee, *Foreign Trade and Tariff Proposals*, pt. 1, p. 242, 255, 313; and *The Financial Times* (February 22, 1965), p. 1.

14. *Congressional Record*, April 28, 1965, p. 8721, 8722; House Ways and Means Committee, *Foreign Trade and Tariff Proposals*, pt. 1, p. 225.

15. This data comes from Masson and Whitely, *Barriers to Trade between Canada and the United States*, p. 64; B. Balassa, *Trade Liberalization among the Industrial Countries*, p. 65; "How Governments Buy," *Industrial Canada* (June 1968), p. 12; *Congressional Record* (May 4, 1965), p. 12.

16. Information on federal policies can be found in Buy-American Act of 1933, 41 *United States Code* 10a to 10d; Executive Order 10582, 10 *Federal Register* 8723, December 17, 1954; Kelly, "Nontariff Barriers," p. 279; Subcommittee on Defense Procurement, *Hearings on the Impact of Military Supply*, p. 357; OECD, Joint Economic Committee, 88th Cong., 1st sess., *Steel Prices, Unit Costs, and Foreign Competition* (Washington, D. C.: United States Government Printing Office, 1963), p. 489.

17. For state Laws, see Hemmindinger, *Non-Tariff Trade Barriers*, p. 11; House Ways and Means Committee, *Foreign Trade and Tariff Proposals*, p. 2902, Texas Association of Steel Importers v. Texas Highway Commission, Docket No. A-9515, Supreme Court of Texas, November 6, 1963; *Newsletter*, West Coast Metal Imports Association, Inc., 3, no. 1 (January 1969).

18. Towle, *International Trade and Commercial Policy*, p. 467.

19. French subsidies are covered in *Time*, (July 5, 1958), p. 56; *European Community*, "Community Deals with French Crisis" (August 1968), p. 3; Senate Finance Committee, *Steel Imports*, p. 39; "ECSC High Authority Decision," *Journal Officials Des Communantes Europeenes*, June 17, 1967.

20. House Ways and Means Committee, *Foreign Trade and Tariff Proposals*, p. 2902.

21. General Agreement on Tariffs and Trade, Article VI, as revised by the Protocol Amending the Preamble and Parts II and III of the General Agreement, dated March 10, 1955; and Balassa, *Trade Liberalization*, p. 67.

22. Information in this section is from House Ways and Means Committee, *Foreign Trade and Tariff Proposals*, p. 2910; and Senate Finance Committee, *Steel Imports*, p. 39.

23. House Ways and Means Committee, *Foreign Trade and Tariff Proposals*, pp. 2002-2004, 2911.

24. The following sources describe the Japanese penalty: Senate Finance Committee, *Steel Imports*, , p. 329, 328; *Japan Metal Trade Bulletin*, January 11, 1968, p. 1.

25. H. G. Johnson, *Economic Policies toward Less Developed Countries*, pp. 27, 82; in 1962, the percentages were 0.66 percent and 1.32 percent respectively. See A. Pincers, "The Cost of Foreign Aid," *Review of Economics and Statistics* 45 (November 1963), p. 364.

26. Properly applied border-tax adjustments do not discriminate against foreign in favor of domestic goods, even if internal taxes are higher in the importing country. Therefore differences between the absolute value of indirect taxes are not considered to erect non-tariff barriers in this study, as they are in other works. See D. Dosser, "Fiscal and Social Barriers to Economic Integration in the Atlantic Area," in B. Balassa, ed., *Studies in Trade Liberalization*, pp. 230, 231; and GATT, *Basic Instruments and Selected Documents*, Geneva, Articles II, III, VI, and XVI.

27. See J. Due, "Introduction: The Issues," *The Role of Direct and Indirect Taxes in the Federal Revenue System*, (Princeton: Princeton University Press, 1964), p. 3.

28. H. Junz, "The Border Tax Issue Defined," *Issues and Objectives of United States Foreign Trade Policy*, Joint Economic Committee, United States Congress, 90th Cong., 1st sess., September 1967, pp. 31-32.

29. OECD, *Border Tax Adjustments and Tax Structures in OECD Member Countries* (Paris: OECD, 1968), p. 17.

30. *Steel Facts*, October, 1963, p. 3.

31. Ibid.

32. See H. Haller, "Taxes: A Universal Problem" *European Community*, (April 1971), p. 7.

33. M. Adelman, "The Corporation Income Tax in the Long Run," *Journal of Political Economy*, December 1967.

34. C. Hall, "Direct Shifting of the Corporation Income Tax," *American Economic Association Papers and Proceedings* (1964), p. 258-71.

35. R. Musgrave and M. Kryzyzniak, *The Shifting of the Corporate Income Tax*, Baltimore: John Hopkins Press, 1963.

36. For instance, see G. Bragg, A. Harberger, and P. Mieszkowski, "Empirical Evidence on the Incidence of the Corporation Income Tax," *Journal of Political Economy*, December 1967.

37. Balassa, *Trade Liberalization*, p. 64.

38. R. Aliber and H. Stein, "Price of United States Exports and the Mix of Direct and Indirect Taxes," *American Economic Review* (September 1963), pp. 703-10.

39. B. Lowery, "The Effect of the Disparate Treatment of Tax Rebates

Upon Trade Patterns of Manufactured Goods under the General Agreement of (sic) Tariffs and Trade,'' Ph. D. thesis, Michigan State University, Graduate School of Business Administration, East Lansing, 1970.

40. GATT *Basic Instruments*, Articles XVI, VI.

41. House Ways and Means Committee, *Foreign Trade and Tariff Proposals*, pp. 2222, 1147.

42. J. Jansen, "Tax Harmonization in the Community," *European Community*, January 1968; and "Commission Sues Italy for Discriminatory Customs Surcharge on Textile Imports," *European Community*, June 1969, p. 70.

43. GATT, *Basic Instruments*, Article VI. Emphasis added.

44. House Ways and Means Committee, *Foreign Trade and Tariff Proposals*, pp. 2913, 2223, 2917; and OECD, *Border Tax Adjustments*, p. 149.

45. Belgian Office of Foreign Trade, *Taux de la Ristorine*, April, 1969, Section 73.

46. OECD, *Border Tax Adjustments*, p. 148, but see, House Ways and Means, Committee, *Foreign Trade and Tariff Proposals*, pp. 281, 3902.

47. House Ways and Means Committee, *Foreign Trade and Tariff Proposals,* p. 2906; letter dated June 25, 1969, United States Embassy, Rome, Italy; Senate Finance Committee, *Steel Imports,*, p. 37.

48. OECD, *Border Tax Adjustments,* p. 37.

49. Senate Finance Committee, *Steel Imports,* p. 37.

50. House Ways and Means Committee, *Foreign Trade and Tariff Proposals* pt. 5, p. 2222.

51. OECD, *Border Tax Adjustments,* p. 36.

52. Dosser, "Fiscal and Social Barriers," p. 230.

53. Refer to Chapter 1.

54. OECD, *Border Tax Adjustments*, p. 36.

55. House Ways and Means Committee, *Foreign Trade and Tariff Proposals*, p. 3881, 3902; *Overseas Business Reports*, 69-30, p. 10.

56. OECD, *Border Tax Adjustments*, pp. 114, 141, 149, 146, 35.

57. House Ways and Means Committee, *Foreign Trade and Tariff Proposals*, pp. 2913-17.

58. Letter dated June 25, 1969, United States Embassy, Rome, Italy.

59. Senate Finance Committee, *Steel Imports*, p. 38.

60. Senate Finance Committee, *Hearings on Senate Resolution No. 149,* Congrsssional Session, Washington, D. C.: United States Government Printing Office (June 3, 1966), p. 273.

61. *Overseas Business Reports*, 64-26, p. 4; House Ways and Means Committee, *Foreign Trade and Tariff Proposals*, pt. 8, p. 3582.

62. "Steel Steps Up the War on Imports," *Business Week* (February *1967), p. 36. Week*

63. *Time* (September 6, 1971), p. 12.

64. House Ways and Means Committee, *Foreign Trade and Tariff Proposals,* pt. 2, p. 642; First National City Bank of New York, *Monthly Economic Letter* (June 1968), p. 1.

65. Masson and English, *Invisible Trade Barriers between Canada and United States*, (Canada: Canadian-American Committee, 1963), pp. 40, 41; *The Canada Gazette Part II,* "Customs Tariff: Surcharge on Imports Order," 96, no. 13 (July 11, 1962), pp. 699-706; First National City Bank, *Monthly Economic Letter.*

66. J. Viner, *Dumping: A Problem in International Trade*, Kelley reprint: New York, 1966, pp. 3, 132, 133. For more on dumping, see Towle, *International Trade*, 667; Kelly, "Nontariff Barriers," p. 296; G. von Haberler, *The Theory of International Trade*, William Hodge and Co., London, 1936, pp. 302-17; Massel, *Competition and Monopoly*, p. 63.

67. GATT, Article VI, *Basic Instruments and Selected Documents*, vol. I, (rev.), Geneva.

68. Kelly, "Nontariff Barriers," . p. 300.

69. Balassa, *Trade Liberalization*, p. 66.

70. Antidumping regulations in the EEC are covered in Kelly, "Nontariff Barriers,", p. 300; GATT, *Antidumping and Countervailing Duties*, Geneva: July 1958, p. 10, 21, 14, 47, 48, 50; *Overseas Business Reports*, 64-31, p. 23; and Hearings before the Subcommittee on Foreign Economic Policy of the Joint Economic Committee, *The Future of United States Foreign Trade Policy*, 90th Cong., 1st sess., vol. I, United States Government Printing Office: Washington, D. C., 1967, p. 15.

71. Canadian antidumping policies are described in *World Trade Information Service*, 62-50, pt. 2, p. 12; Kelly, "Nontariff Barriers," p. 297; GATT, Antidumping, pp. 53, 10.

72. Masson and Whitely, *Barriers to Trade between Canada and the United States*, p. 53.

73. *The Future of United States Foreign Trade Policy*, 15.

74. GATT, Antidumping, pp. 5, 6.

75. *Overseas Business Reports*, 67-91, pp. 5,6.

76. *International Commerce* (April 6, 1968), p. 24.

77. "European Non-Tariff Obstacles Critizcized," *European Community* (August, 1965), p. 8.

78. GATT, Antidumping, p. 6.

79. British practices are covered in *Overseas Business Reports*, 65-11, p. 6; GATT pp. 5, 6; *Overseas Business Reports*, 68-38, p. 4; *Hearings on Senate Resolution 149*, p. 342; Kelly, "Nontariff Barriers,", p. 300.

80. Data in this paragraph came from "Europeans Weigh United States Non-Tariff Obstacles to Trade," *European Community* (July 1965), p. 9; Kelly, "Nontariff Barriers," p. 297; GATT, Antidumping, p. 10; Tariff Act of 1930, Section 303, 19 U. S. C. 1303; *The Future of United States Foreign Trade Policy*, p. 15.

81. GATT, Antidumping, p. 117.

82. Masson and Whitely, *Barriers to Trade*, p. 35.

83. United States Tariff Commission, "Hot-Rolled Carbon Steel Wire Rods from Belgium" Tariff Commission Publication 93, Washington, D. C., Luxembourg, West Germany, 1963, France 94, 95, 99.

84. *The Iron Age*, January 11, 1973, p. 27; February 1, 1973, p. 27; February 8, 1973, p. 75.

6 Administrative Practices

Although intended to implement other import restrictions such as tariffs and quotas, customs administration in itself may erect barriers that are more difficult to surmount than the original restrictions. The emphasis which the Common Market has placed on achieving complete harmonization of its member's customs procedures provides one indication of their importance.[1] The action of the United States Congress, which was so impressed by the protective effects of certain valuation procedures that in enacting changes it specifically exempted certain goods from new and simpler methods, also reflects the importance of administrative practices.[2] For all that, however, we will find the restrictive effects of this type of barrier more difficult to quantify, except in the cases of deposits, fees, or terms of duty-payment.

This chapter describes each country's customs practices in the order they occur in the importation process. Before he can sell his goods, an exporter must enter his advertising and samples for customs scrutiny, but none of the major steel-trading nations restrict steel in this way. Next, he must obtain information on customs procedures. Most industrial countries give this freely, and only steel importers in the United States have found incomplete information to be an important impediment. The uncertainty of changes in customs regulations and delays in the processing of imports also may surround customs administration. United States steel importers list uncertain changes as important barriers, while delays due to customs complexities have been found in both the United States and Japan by steel traders.

Import Marking Requirements

Prior to shipment the exporter must make sure that his goods satisfy customs marking requirements. Marks of origin may have the intent of preventing consumers from being misled by informing them of the origin of their purchases. However, the effect of marking requirements may extend beyond consumer protection and erect import barriers which protect domestic producers. Implicit in mark-of-origin rules is the assumption that, if domestic consumers were apprised of the good's foreign origin, they might show a preference for domestic substitutes. On the supply side, the cost of marking may discourage exports from foreign countries.

With the exception of specific marking rules for a few items, principally foodstuffs and drugs, the countries considered here usually confine their import requirements to prohibiting marks which indicate a false domestic origin. Indeed, the United States is the only major steel trader which imposes a general marking requirement on imports.[3] Although the Treasury Department may grant exceptions to the marking rule where the cost is "economically prohibitive," steel receives no exemption. The Customs Bureau publishes details of decisions which define the type of marking necessary for each good to meet the Tariff Act's general specifications. The act requires imports to be marked (1) in a conspicuous place, (2) legibly, indelibly, and permanently, and (3) with the English name of the country of origin. Improperly marked goods may not always be remarked and imported. Instead, they may be reexported, destroyed, or subject to fine.

United States marking requirements restrict imports by creating uncertainty and raising the cost of imports. The chief problem of uncertainty appears to stem from inadequate information from customs on what will be accepted as a proper designation or method of marking. The uncertainty over disposition of improperly marked goods creates another difficulty. Although the Treasury can waive marking which entails prohibitive costs, it has not exempted such products as watch bearings and hypodermic needles. The marking of iron and steel products is reported to involve the expensive process of countersinking.[4] United States importers of French structural steel have complained that the expense of marking each section with the word "France" has presented an insurmountable obstacle to this type of import.[5]

Documentary Requirements

Among the major steel-trading countries, those belonging to the Common Market have similar documentary requirements. We can say the same for all the major steel traders as well, but the degree of similarity is less. All ten countries ask for a number of documents so that differences in restrictiveness depend only upon differences in the number and specification of the papers. In general, governments require only bills of lading and/or commercial invoices, declaration forms, and certificates of origin. However, in Belgium and Luxembourg a second declaration must be made; Canadian importers must prepare invoices in triplicate and must give other information specified by customs; French customs may ask for additional documents such as contracts or correspondence; a complex declaration form and an import control report are unique German restrictions. In general,

German customs officers are reported to be lenient with regard to any necessary correction of documents. Requests for advance rulings on Italian customs classification sometimes must appear on special paper which costs about $0.34 a sheet. Although British customs require a variety of declaration forms, they do not have stringent rules regarding the content of other import documents.[6]

World steel traders rank United States documentary requirements eighth in a field of twelve in terms of restrictiveness. Although little uncertainty exists over the United States requirements since information and forms are freely available from United States consulates, the requirements are more complex than in most other major steel-trading countries. In contrast to most countries, which require no particular form, United States importers must present a special customs invoice for each shipment to the United States. For the few items exempted from the special invoice requirement, America requires a commercial invoice for customs purposes. If an invoice cannot be produced, then the importer must fill out a pro forma invoice and post a bond guaranteeing that an invoice will be presented within six months.[7]

The United States eases the burden of its documentary requirements by allowing invoices to be filled out in any language. On the other hand, the United States customs ask for original invoices and other papers to support invoices which cover single shipments of more than one type of merchandise. Moreover, their invoices require a greater amount of information than those of most foreign countries. The list of invoice items includes the port of entry, time and place of sale, name of buyer and seller; name, quality, description, marking, quantity, price, currency of purchase, delivery charges, rebates, subsidies, current home-consumption price, value, place of origin, and any other information necessary for valuation or classification purposes.

According to the United States Secretary of Transportation,[8] an average overseas shipment requires more than 360 copies of 46 separate documents. As a result American exports and imports alone create 6.5 billion copies a year of 828 million documents. These estimates include papers for banks, insurance companies, and shipping firms, but most go to government agencies. The cost accounted for 7.5 percent of the value of United States export and import shipments, and 45 percent of the cost could be regarded as unnecessary. Assuming that these aggregate American figures apply equally well to the steel imports of all countries, we can use 3.37 percent as the ad valorem rate of this nontariff barrier.

This rate will not apply precisely to all the major steel traders, but it represents a suitable first approximation. Based on requirements concerning the submission time, amount, and content of documents, we can rank countries according to the restrictiveness of their documentary regulations: France, Canada, United States, Germany, Japan, Italy, Benelux, United Kingdom. However, a more precise appraisal of the restrictiveness of this barrier would require details of the cost of preparing documents for each shipment. Obviously the cost per unit would be small for large shipments, and for large firms there may be economies of scale in retaining a full-time customs lawyer and broker. With regard to small companies, however, congressional studies report that many do not participate fully in world trade, at least partly because of the complex and burdensome requirements for documentation.[9] The present study's survey of United States steel producers reaffirmed this observation. Replies from small companies state that they did not export simply because of the complexities involved.

Customs Classification

Customs rules relating to valuation and classification can be regarded as separate trade barriers when they require lengthy and costly litigation or create uncertainty. Because these restrictions relate directly to tariffs, they are sometimes referred to as paratariff barriers.[10] This section reviews the major steel-trading countries' customs classification procedures, under which the respective governments assign a particular rate of duty according to a tariff schedule.

All Common Market members, as well as the United Kingdom and Japan, subscribe to the Brussels Tariff Nomenclature (BTN), though this does not necessarily mean that they have the same tariff schedule classifications, for they often use different subdivisions under the same general headings. Nevertheless, these countries belong to the Customs Cooperation Council, a group established to maintain uniformity in the application of the BTN. Thus these countries manage to greatly reduce uncertainty and other problems concerning import classification.

The Canadian classification system differs markedly from the BTN. The system is comprehensive in that every import must be classified somewhere in the schedule; if the schedule lists a product under more than one category, the one with the highest rate of duty shall apply. As a result of imports entering at more than one rate of duty, the Canadian customs reserves the right to make additional charges or refunds without any time limit. The year 1951, for instance, saw 120,405 such adjustments, of which there were five additional duties for each refund.

Canadian customs classification makes another distinction between imports of a class or kind made in Canada and imports with no domestically produced substitutes. The ambiguity of the words "class or kind," which may refer to either a narrow or wide range of products, leaves a good deal of discretion in the hands of customs officials. Moreover, a product is "made in Canada" only if 10 percent of domestic consumption consists of Canadian goods, a situation that can change from year to year. Only three weeks notice of change in this classification is necessary.

Further complexities in Canadian import classifications arise over the final use of the import, a cause for higher or lower duties. Although this distinction has contributed to much uncertainty and delay, primary iron and steel mill products at least benefit now from a 1958 overhaul of their tariff section which eliminated many end use items.

The American Tariff Act of 1930 contained 730 main classifications of imports, which were broken down further into numerous subclassifications. In contrast to Canada, the United States has fewer classification disputes (50,000 pending in 1953) than protests over valuation. The United States does not forestall entrance of imports during classification disputes as it does in disagreements about valuation.[11]

One of the main problems of customs classification in the United States stems from the rule of similitude, which attaches the tariff rate of similar items to those not specifically enumerated in the tariff schedule. As in Canada, if the article could be included in more than one classification, customs places it in the one with the highest duty. Thus new products cannot be imported duty-free even when they more closely resemble goods on the free list than those on the duty list. In addition, the law admonishes customs officials to find the most restricted product which resembles the new import. Another problem stems from classifying imports according to their chief component by value: customs may charge different duties for the same items, ashtrays for instance, because one is made of metal and the other is ceramic. However, the largest share of these classification restrictions appears to fall on finished manufactures, and not on primary goods like steel.

The *Tariff Schedules of the United States*, a major revision of the classification system, became effective in September 1963. The consolidation and rearrangement of the old schedule has eliminated many problems of complexity. In addition, the inclusion of many new products has lessened the importance of problems arising from application of the rule of similitude and the chief-component-by-value standard.[12] Despite this simplification, other countries still find reason to complain about the "very complicated headings" in the tariff schedule and "arbitrary changes of classification."[13] We can more objectively appraise the effect of the many customs simplification attempts of the United States by noting that the number of tariff classification cases in court over the years has not declined, but increased. Including the number of those decided in each year also indicates the increase in delays involved in such cases, since both the relative and absolute number of cases decided each year has declined. The number of cases by year is presented in Table 6-1.

The customs bureaus of all ten major steel-trading countries give advance rulings on import classifications upon receipt of a request and adequate information, but the United Kingdom and Germany will not give definite rulings until the time of actual importation. Italy requires detailed requests written in Italian, while both France and Italy require special application forms for advisory rulings.

It should be noted that Canadian tariff quotations cannot be relied upon since they can change without notice. In contrast, notice of the United States Custom Bureau's intent to alter a classification is published in the *Federal Register* 120 days prior to the actual change.

In general, appeals of classification rulings can be made first to the customs bureau and eventually to the courts, but in France and Japan classification disputes cannot be appealed to the judiciary. In Canada, where protests over customs classification are the most common form of import litigation, appeals may not be undertaken on an adverse indication of classification, but only on a classification as applied to an actual import. As a consequence, the higher duty imposed by an erroneous advance ruling may prevent not only importation, but the possibility of correcting the mistake as well.[14] During litigation, duties must be paid or the goods cannot be imported; some countries (Benelux members and Japan in particular) will refund excess duties on goods imported before a successful appeal.

Valuation

Among the major steel-trading countries considered here, members of the EEC and the United Kingdom are signatories to the "Convention on the Valuation of Goods for Custom Purposes," most commonly known as the Brussels Definition.[15] Although not a signatory of the valuation convention, Japan also had adopted the Brussels Definition in its valuation practices.[16] Accordingly, the countries use a "normal price" as the base to which they apply ad valorem duties. The "normal price" equals that which imports "would fetch at the time when the duty becomes payable on a sale in the open market between buyer and seller independent of each other.[17] If a transaction meets these requirements then the "normal price" corresponds to the c.i.f. invoice price. In practice, signatory countries use the invoice price unless it is lower than the "normal price."[18] Importing countries exclude duties and taxes from "normal price" but all other charges for delivery to the place of clearance are included.[19]

Table 6-1
United States Customs Court Classification Protest Cases

Fiscal Year	*Pending during Year*	*Decided during Year*
1958	161353	38478
1959	156704	25094
1960	164007	59336
1961	141883	47098
1962	117711	35479
1963	104330	17111
1964	113674	27784
1965	109645	19497
1966	NA	NA
1967	204347	27908
1968	261921	33528
1969	290096	32105
1970	311873	41643
1971	304166	17067

Source: Director of the Administrative Office of the United States Courts, *Annual Report, 1959, 1961, 1962, 1963, 1964, 1965, 1968, 1972* Washington, D.C.: United States Government Printing Office.

The Brussels Definition may impede trade when the condition of independence between buyer and seller does not hold. If the importer is the exclusive agent, sole distributor, branch, or subsidiary of a foreign firm, the invoice price may be lower than the "normal price" by the amount of discounts for advertising and other selling expenses. To offset this difference, European customs authorities impose a 1 to 10 percent increase in the invoice price. The amount of increase is sometimes negotiated between customs officials and importers, and it is particularly arbitrary in cases where the related firm is the sole buyer, so that the customs appraiser has no other transactions to guide his determination of normal price.[20]

Canada uses a valuation method similar to the Brussels Definition for about three-quarters of its imports, but the rest appear to permit arbitrary valuation. The Canadian "similar goods" rule, for instance, states that the value of some imports for duty purposes is the cost of production of the import plus some gross profit markup. Canada also employs a discretionary valuation rule, whereby the Minister of National Revenue can prescribe the manner in which value is determined for imports such as those from state monopolies or those intended for further processing.[21] In addition, a valuation rule similar to the United States escape clause permits the minister to increase the dutiable value whenever imports materially injure Canadian industry. Finally, another rule permits the minister to set dutiable value at the average of previous periods, if he finds that import prices are below normal.[22] With five methods of valuation, four of which permit wide ministerial discretion in valuing imports, it is not surprising that Canada has been the object of foreign complaints concerning arbitrary valuation practices.[23]

United States valuation practices also differ from the procedure specified by the Brussels Definition. Four customs valuation methods are expressly prescribed in the American Customs Simplification Act of 1956, but the law allows no administrative discretion to be exercised in valuation, as it is in Canada. The first duty base is *export value*, a term defined much like the Brussells Definition's normal price except that the export value is f.o.b.[24] According to a 1961 study, the United States appraised 87 percent of all import invoices on the basis of export value.[25] If the export value cannot be determined satisfactorily, customs refers to the dutiable amount as *United States value*, the wholesale price of the imported product in the United States minus commissions, profits, customs duty, transportation cost, insurance, and other delivery expenses. United States value is usually higher than export value probably would have been, because the law sets the maximum allowance for profits at 8 percent and for commissions at only 6 percent, "in each case well below average."[26] A third method values imports according to the *constructed value*, which is the estimated cost of production in the exporting country.[27] These three methods of valuation have been subject to European complaints about their complexity, and uncertainty, but foreign criticism has stopped short of calling United States procedures arbitrary.[28]

The fourth United States valuation rule, American-selling price valuation, has been roundly criticized by Europeans, because the rule conflicts with the GATT provisions that customs value should not be based on the value of similar goods produced in the importing country.[29] American-selling price valuation has not been applied to steel, however, and in fact the duty base for most United States steel imports is export value.

The United States Congress exempted 1,105 tariff items whose dutiable value would have been reduced by 5 percent under the new valuation provision of the Customs Simplification Act of 1956.[30] But no iron and steel mill products were among these "final-list" items.[31]

Apart from complex and arbitrary valuation methods, considerable delays in the customs appraisal process often occur. Unfortunately, little information is available on delays or the reasons for them, except in the United States. As early as 1938 United States importers complained that "invoices have been witheld for as long a period as three years, and it is not uncommon for such returns to be delayed for six or eight months or a year."[32] In 1953, over 16 percent of all invoices on hand remained in the United States Customs appraiser's offices over two years, and almost one-fourth were there from one to two years.[33] Between 1952 and 1958, the number of invoices on hand over ninety days nearly doubled.[34] Valuation protests involve further delay as the small proportion of cases decided each year indicates. Table 6-2 shows that customs court reached a decision in only 5 percent of the pending cases in 1968. Moreover, 1968 appeal decisions took an average of fifteen months, as shown in Table 6-3. These and other valuation problems can affect any import, including iron and steel mill products, but their restrictive effect is impossible to quantify.

Table 6-2
United States Customs Court Appeals for Reappraisement of Customs Valuation

Fiscal Year	*Pending during Year*	*Decided during Year*
1958	102696	16028
1959	109044	20663
1960	108128	32276
1961	102716	12858
1962	107724	13855
1963	107268	8144
1964	122466	10166
1965	136651	8921
1966	NA	NA
1967	174212	10619
1968	185758	9867
1969	189652	16428
1970	189914	23996
1971	170916	15995

Source: Same as Table 6-1.

Table 6-3
United States Court of Customs Appeals: Average Number of Months between Filing and Decision

1958	8.0
1959	9.8
1960	10.8
1961	10.1
1962	9.7
1963	9.0
1964	11.4
1965	13.0
1966	16.6
1967	14.0
1968	15.2

Source: Same as Table 6-1.

Customs Fees and Deposits

Among the many important restrictive practices associated with customs administration, only handling fees, statistical charges, documentary expenses, and customs deposits appear susceptible to quantitative analysis. However, the fact that countries cannot disguise these barriers in the respectable cloak worn by other administrative practices, may contribute to their infrequent use.

France subjects imports to a customs stamp tax of two percent of all customs charges except the value-added tax,[35] which is the equivalent of 0.2 percent of the c.i.f. import price. In addition, France levies an administrative fee of 0.2 percent of all border charges,[36] which equals 0.06 percent of the import price.

Italy presents a slightly smaller customs restriction in the form of an administrative fee and a statistical tax. The administrative charge equals 0.5 percent of the c.i.f. import price.[37] The statistical tax has the specific rate of Lire 10 per quintal, equivalent to 0.1 percent of 1969 Italian steel import prices.[38]

In Japan an import license applicant usually must deposit from 1 to 35 percent of the c.i.f. import value with a foreign exchange bank until the goods clear customs.[39] The customs deposit rate equals 1 percent for pig iron, steel ingot, and semifinished steel products and 5 percent for all other steel mill products.[40] Japan's apparent home consumption of ingots, pig iron, and semifinished steel products amounted to 50 percent of total 1966 consumption with shipments of finished iron and steel mill products making up the balance.[41] Weighting the import deposits by domestic consumption shares gives us an average deposit requirement of 3 percent. Six months appears to be the average length of time that customs holds the import deposit. Taking the 1969 Japanese interest rate of 6 percent as the cost of capital tied up in import deposits,[42] the cost of providing the deposit as a percentage of the c.i.f. import price amounts to 0.09 percent. In comparison, the cost of the 35 percent maximum deposit in Japanese imports as a percentage of price equals 1.05 percent.

The United Kingdom also maintains an import deposit plan that requires most British importers to keep much of their capital tied up.[43] The scheme applies to all manufactured goods, including steel, but excludes food stuffs, fuel, raw materials, and any goods from developing countries. The deposit must be paid before imported goods are cleared through customs, and no refunds occur for six months. The amount of the import deposit is equal to 50 percent of the c.i.f. import value.[44] At the 1969 British interest rate of 7.5 percent, the cost of the deposit as a percentage of price amounts to 1.9 percent.[45] United States steel exporters found this barrier to be the fourth most restrictive in Britain.

Other Customs Provisions

The only administrative practices, besides penalties, that have not been covered so far mainly relate to the movement of imports through customs. Information concerning the clearance, handling, and storage of goods by customs in all countries can be found in a variety of United States government publications.[46] Since the procedures generally resemble each other, separate consideration of each country's customs procedures will not take place. Instead, the physical processing of imports will be described in sequence, with exceptional practices being noted at each step.[47]

Upon arrival of imports in any of the countries considered here, they must be formally entered with the customs within the time limits listed in Table 6-4.

Imports not entered with the customs may remain in the customs warehouse for one year, Italy being an exception with a three-month limit. All applicable storage and handling fees, as well as risk of loss, are assigned to the importer. Unclaimed imports are subject to the same treatment as abandoned goods.

Abandonment of imports may occur for a variety of reasons, among them spoilage, damage, or the importer changing his mind. They will be held for a time as discussed above, and then be sold, destroyed, or reexported. All expenses, taxes, and duties must be paid by the importer in the case of destruction or the exporter in case of reexportation. Proceeds from the sale of abandoned goods first cover import liabilities, and then at least in the United States, Canada, Belgium, Luxembourg, and Italy, the proceeds go to the owner of the goods. France will not take over abandoned goods, if the expense of doing so is high. Japan and the United Kingdom apparently make no attempt to refund proceeds net of expenses to the owner of the goods.

Entering goods with customs usually entails depositing them in a customs warehouse until duties are assessed, except in Japan and the Netherlands. After entry, importers who do not wish to use the merchandise

Table 6-4
Days Allowed for Customs Entry after Arrival of Imports

Belgium	10
France	1
West Germany	7
Japan	15
Luxembourg	10
United States	5

Sources: See note 48.

immediately may postpone their payment of duty, though the United States and Canada may require deposit of the estimated duty. Until the duty is paid, imports usually must be kept under bond or some other type of security in customs-approved public or private warehouses, where repacking, marking, manufacturing, or other manipulation of imports is usually permitted, except in France, Italy, and the United Kingdom.

Most of the major steel-trading countries permit imports to remain in storage warehouses about three years, though Canada only permits them to stay two years, while Germany allows five. The varying terms and conditions applying to storage have been recognized as important nontariff trade barriers.[48] Some countries, like the Netherlands, provide customs services free of charge; others, such as the United States, levy fees for any extra customs supervision. Some countries charge higher rent on warehouses, permit internal taxation of stored goods which have not been officially imported,[49] or charge the highest duty when tariffs change during the storage period.[50]

When a country levies duties, whether at the time of arrival or after some storage period, the method of payment may also vary from country to country. Prepayment of duty by the foreign exporter is either prohibited or raises the valuation of the imports. Most countries require all customs charges to be paid in domestic currency. Perhaps most important to importers are the deferred payment provisions of Europe. In Italy, importers must pay duties when they take possession of goods, but the Benelux countries allow importers to postpone payment for a week. Some EEC members give imports three months' credit free of interest, while others charge interest, but allow installment payments on duty charges.[51]

Besides administrative fees and other customs charges involving the treatment of imports, customs may also impose certain penalties for violation of its regulations. The major steel traders sanction forced sales, fines, confiscation of goods, the means of transport, and imprisonment. The penalties vary from country to country, but do not appear to have been imposed arbitrarily. Moreover, there is no evidence to suggest that steel has been restricted by these penalties.

Notes

1. See *European Community* (November 1968), p. 18; and (June 1968), p. 4.

2. Masson and Whitely, *Barriers to Trade*, p. 29.

3. For information on American marking requirements, see Tariff Act of 1930, Section 304 as amended, 19 USCA, par. 1304 (1960); and publications of the United States Treasury Department, Bureau of Customs:

Exporting to the United States (Washington, D. C.: United States Government Printing Office, pp. 29-33, *Digest of Decisions for the Marking of Imports* (Washington, D. C.: United States Government Printing Office, 1968),; *Marking of Country of Origin on United States Imports* (Washington, D. C.: United States Government Printing Office, 1967).

4. Masson and Whitely, *Barriers to Trade*, p. 33.

5. Commission on Foreign Economic Policy, *Staff Papers*, (Washington, D. C.: United States Government Printing Office, 1954), p. 355.

6. *Overseas Business Reports*, 68-30, p. 6; and 68-40, p. 3.

7. Documentary requirements are summarized in United State's Treasury Department, *Exporting to the United States,* pp. 8, 18-20; and United States Treasury Department, Bureau of Customs, *United States Customs Regulations*, Section 8.15.

8. As reported in *Time* (December 6, 1971), p. 94.

9. House Ways and Means Committee, *Foreign Trade and Tariff Proposals*, pt. 7, p. 3110.

10. Kelly, "Nontariff Barriers," pp. 284, 285 is the primary source on classification. See also Masson and Whitely, *Barriers to Trade*, pp. 48-50, 56, and Masson and English, *Invisible Trade between Canada and the United States*, Canadian-American Committee of the National Planning Associaion, (United States) and Private Planning Association of Canada, pp. 19, 20.

11. American classification is described by Commission on Foreign Economic Policy, *Staff Papers* (Washington, D. C.: United States Government Printing Office, 1954), pp. 331, 333, 335, 343-358.

12. United States Tariff Commission, *Tariff Classification Study,* Washington, D. C.: United States Government Printing Office, 1960.

13. *European Community*, (July 1965), p. 8.

14. Masson and English, *Invisible Trade*, p. 18.

15. Kelly, "Nontariff Barriers," p. 285.

16. Ibid.

17. Customs Cooperation Council, *The Brussels Definition of Value for Customs Purposes* (Brussels: April 1964), p. 41.

18. *Overseas Business Reports*, 67-91, p. 2.

19. International Chamber of Commerce, *Customs Valuation of Imported Goods,* February 1959.

20. International Chamber of Commerce, *The Brussels Definition of Value: The Case of the Sole Buyer* (February 1963), p. 5.

21. Masson and Whitely, *Barriers to Trade*, p. 53.

22. Kelly, "Nontariff Barriers," p. 294.

23. House Ways and Means Committee, *Foreign Trade and Tariff Proposals,* pt. 1, p. 229; pt. 9, p. 3923.

24. American Importers Association, *United States Customs Valuation Procedure* (New York: 1958), p. 5.

25. United States Treasury Department, Bureau of Customs, *An Evaluation of Mission: Organization, Management* (December 1964), pt. VI, p. 24.

26. Towle, *International Trade and Commercial Policy,* p. 537.

27. Kelly, "Nontariff Barriers," p. 288.

28. *European Community* (July 1965), p. 8.

29. GATT, Article VII.

30. Kelly, "Nontariff Barriers," p. 288.

31. United States Treasury Department, *Exporting to the United States* (March 1965), pp. 65-77.

32. Bidwell, *The Invisible Tariff,* p. 34.

33. Commission on Foreign Economic Policy, *Staff Papers*, p. 338.

34. Masson and Whitely, *Barriers to Trade*, p. 28.

35. *Overseas Business Reports*, 67-91, p. 2.

36. Senate Finance Committee, *Steel Imports,* p. 316.

37. *Overseas Business Reports*, 68-40, p. 6.

38. OECD, *The Iron and Steel Industry,* Tables 33 and 34.

39. United States Department of Commerce, "Japans Licensing and Exchange Controls," *Overseas Business Reports*, 62-2 (Washington, D. C.: United States Government Printing Office, November 1962), p. 1.

40. Letter dated July 14, 1969, United States Embassy, Tokyo, Japan, p. 2.

41. Crude steel consumption: OECD, *The Iron and Steel Industry,* Table 27; and domestic shipments of finished steel: Japan Iron and Steel Federation, *JISF Statistical Yearbook for 1966,* Table 44A.

42. International Monetary Fund, *International Financial Statistics* (May 1969), p. 185.

43. *The Wall Street Journal* (March 3, 1969), p. 1.

44. International Monetary Fund, *International Financial News Survey* (December 6, 1968), p. 405.

45. Based on Bank Rate, International Monetary Fund, *International Financial Statistics* (May 1969), p. 319.

46. For example, in the case of Japan see: United States Department of Commerce, "Japan's Licensing and Exchange Controls," *Overseas Business Reports*, 62-2; United States Department of Commerce, "Japan Lifts

Import Restrictions on 39 Additional Commodities,'' *International Commerce* (Washington, D. C.: United States Government Printing Office, October 26, 1964), p. 14; United States Department of Commerce, ''Preparing Shipments to Japan,'' *Operation Reports* (Washington, D. C.: United States Government Printing Office, 1960), pt. 2, No. 60-9; United States Department of Commerce, ''Foreign Trade Regulations of Japan,'' *Overseas Business Reports*, 65-11.

47. Sources used in this section are the same as those in note 46; United States Department of Commerce, ''Foreign Trade Regulations of Italy,'' *Overseas Business Reports*, 68-40; ''Foreign Trade Regulations of France,'' *Overseas Business Reports*, 67-91; ''Foreign Trade Regulations of Netherlands,'' *Overseas Business Reports*, 64-37; ''Foreign Trade Regulations of the Republic of Germany,'' *Overseas Business Reports*, 68-30, 1968; ''Foreign Trade Regulations of the United Kingdom,'' *Overseas Business Reports*, 68-83; ''Foreign Trade Regulation of Belgium-Luxembourg,'' *Overseas Business Reports*, 64-26; ''Foreign Trade Regulations of Canada,'' *World Trade Information Service*, pt. 2, 62-50, 1962; United States Treasury Department, *Exporting to the United States*, 1965; and United States Treasury Department, Bureau of Customs, *United States Import Requirements*, Washington, D. C.: United States Government Printing Office, 1965.

48. See *European Community* (July 1965), p. 8.

49. In the United States, imports are supposed to be exempt from property tax until sold domestically. See L. Baker, ''When Is an Importer Not an Importer,'' West Coast Metal Importers Association, *Newsletter* (November 1968), p. 3. Nevertheless, before being stopped by the Michigan Supreme Court, the city of Detroit was assessing one million dollars a year on imported steel in warehouses. See West Coast Metal Importers Association, *Newsletter* (January 1969), p. 2.

50. *European Community* (June 1968), p. 7.

51. Ibid.

7 Internal Policies and Regulations

The domestic policies of any government often affect its international trade, whether they intend to or not. We can usefully classify these policies by intent for purposes of exposition, despite the arbitrariness of such a division. This section also divides internal policies according to their purpose. We consider economic policies first, including domestic subsidization, indirect tax policy, price controls, and patent laws. Other internal policies considered are technical specifications and national security restrictions. Steel avoids such restrictions as special-use taxes, credit controls, advertising restrictions, trademark and copyright laws, and health and safety standards. Internal direct tax policies, moreover, do not appear to restrict any imports, including steel.

Internal Economic Policies

Domestic Subsidies

A domestic subsidy which increases domestic supply will also cause a downward shift in the corresponding import demand function. The reduction in import demand facing foreign suppliers restricts imports just as if it had been caused by a tariff.[1] However, the domestic subsidy only decreases the demand for foreign goods to the extent that it is passed on to domestic buyers. Whenever domestic demand and supply elasticities are such that the subsidy is not completely shifted forward, the restrictive effect of the subsidy will be less than that of a tariff of equal amount.

Whenever quantifiable subsidies granted on the basis of output are found below, they are expressed at an ad valorem rate. In the cases of lump-sum subsidies or reductions in fixed costs, however, we assume that they are not passed on in the form of lower prices. While this study makes note of such subsidies, no attempt is made to express them at an ad valorem rate, since they do not affect short-run domestic supply.

Direct Domestic Subsidies. Over the 1947-50 period, Japan paid a subsidy for each unit of domestic steel output with the specific intent of lowering prices for steel users.[2] In the late 1960s, however, only two countries among the major steel traders granted substantial direct cash subsidies to their steel industries. Belgium rebates up to 30 percent of the cost of any industrial buildings and 10 percent of the cost of any industrial equipment located in specified development regions.[3]

In the United Kingdom, the treasury has covered the 1968 British Steel Corporation deficit of $29 million.[4] This subsidy appears to be necessary because the nationalized British industry must give "loyalty rebates" to domestic buyers who purchase no foreign steel over a six-month period.[5] These rebates may run as high as $11 per ton.[6] Taking this upper limit as a percentage of the 1968 domestic price on a high-priced product like cold-rolled sheets ($136)[7] the rebate is estimated to be about 8 percent. Since this is the amount actually passed forward to steel buyers, the import must cost at least 8 percent less in order to remain competitive after the rebate. Therefore, we look upon 8 percent as the ad valorem rate of this nontariff barrier.

Depreciation. Depreciation defines an accounting procedure which distributes the cost of tangible capital assets over their estimated useful life. As an expense, depreciation reduces profits and thus taxes in each year. Straight-line depreciation spreads the total cost of an asset equally over all periods of its life. Accelerated depreciation allows larger charges during the early life of a fixed asset than during later years. Although accelerated depreciation probably does not reflect the actual rate at which an asset loses its productive value, many countries sanction this accounting method for income tax purposes. Moreover, the United States steel industry claims that government subsidization of foreign steel industries by this method has placed the Americans at a competitive disadvantage.[8]

Accelerated depreciation may involve a certain amount of subsidization. For new or expanding business in particular, and for all firms to a certain extent, deferring income taxes is attractive, since it may permit an earlier retirement of debt created to finance the purchase of the assets being depreciated, or may permit the use of such cash for additional equipment. The amount of subsidization involved in accelerated depreciation equals the interest saved through the use of funds destined to pay future taxes. While this calculation is conceptually neat, it is empirically unmanageable, because we cannot find the original value and age of most capital assets. Nevertheless, a brief survey of the differences in depreciation procedures in the major steel-trading countries indicates that the United States does have relatively low depreciation rates, not only for industrial buildings and equipment in general, but for steel plant and machinery in particular. We obtain this information, summarized in Table 7-1, from the Office of Financial Analysis and Director of International Tax Affairs of the United States Treasury. However, we must regard even the maximum and minimum rates shown as rough averages from which a considerable amount of

Table 7-1
First-Year Depreciation Allowances[a]

	Industrial Equipment			*Industrial Buildings*		
	Average Steel	*Maximum[b]*	*Minimum[c]*	*Maximum[b]*	*Minimum[c]*	*Average Steel*
Belgium	NA	20.0	10.0	6.0	3.0	NA
Canada	45.0	100.0	6.0	15.0	5.0	15.0
France	NA	33.0	5.0	12.5	5.0	NA
West Germany	20.0	20.0	10.0	3.5	1.5	3.5
Italy	35.0	49.0	3.0	23.3	5.0	23.3
Japan	33.5	33.3	2.5	20.0	2.0	20.0
Luxembourg	NA	16.0	2.0	3.0	2.0	NA
Netherlands	NA	10.0	7.0	6.0	2.0	NA
United Kingdom	19.0	19.0	2.5	24.0	4.0	24.0
United States	20.0[d]	20.0	10.0	4.4	2.2	4.3

[a]As a percentage of original asset value.

[b]Under accelerated.

[c]Under straight-line depreciation.

[d]Maximum.

Sources: Joint Economic Committee, *Steel Prices*, pp. 434-73; Senate Finance Committee *Steel Imports*, pp. 43, 44; United States Internal Revenue Service, *Internal Revenue Code*, Section 179; *Revenue Procedures*, 62-21, 1962-2 CB 418.

dispersion might be expected, since many countries do not provide statutory asset lives or rates.

Another subsidy arises from the practice of permitting more than the difference between the original value and the salvage value of an asset to be depreciated. This is standard practice among Common Market countries, with the exception of the Netherlands. In Belgium and France, rates of depreciation are set by the tax authorities for each company individually, a practice which leaves the door open to subsidization. The Japanese also allow asset revaluation in an open attempt to subsidize their steel industry.[9]

According to traditional economic theory, the effect of a change in fixed costs does not influence the marginal cost schedule which is the short-run supply curve of a competitive firm. Since the saving from accelerated depreciation is on the cost of a fixed asset, in the short-run at least, such a subsidy seldom benefits consumers. Therefore depreciation subsidies may not erect any nontariff barriers to imports.

Investment Incentives. Another domestic subsidy comes in the form of tax exemptions to firms which expand or modernize. Table 7-2 summarizes the subsidization rates in the major steel-trading countries. Once again, the

Table 7-2
Investment Incentives

	Plant	*Equipment*
Belgium	5-year real estate tax exemption	10%/yr. deduction for 3 yrs.
Canada	None	None
France	Increases in depreciation	Increases in depreciation
West Germany	None	None
Italy	10%/yr. for four years	None
Japan	10% tax credit and write-offs of ⅓ to ½ in excess of depreciation	10% tax credit and write-offs of ⅓ to ½ in excess of depreciation
Luxembourg	10-30% write-offs	10-30% write-offs
Netherlands	5%/yr. for 2 years	5%/yr. for 2 years
United Kingdom	None	25-45%
United States	7% tax credit	7% tax credit

Sources: Joint Economic Committee, *Steel Prices*, pp. 45, 193, 435, 462, 463, 471, 472. Senate Finance Committee, *Steel Imports*, pp. 45, 46, 47.

United States relies relatively less on this type of subsidization, though Canada and Germany do not use it at all.

Low-Cost Loans. A common method of domestic subsidization takes the form of low-cost government loans or preferential interest rates on bank borrowing. While it has been reported that "many foreign steel industries have received government loans at significantly reduced interest rates,"[10] specific evidence has been obtained for only three of the major steel-trading countries of the world.

The French government made two recent loans to the domestic steel industry, totalling 3.3 billion francs; one allows no payment or interest charges for five years, and both loans have interest payments at half the current market rate. In Germany, government loans have prevented serious financial deterioration in the domestic steel industry.[11]

The Japanese Ministry of International Trade and Industry directs the allocation of capital to specific industries.[12] The American steel industry claims that Japanese government control over financial institutions affects the credit availability and terms granted the domestic steel industry.[13] This claim is substantiated by a congressional study which found that "interest-free government loans have been extended in some countries such as Japan."[14] On the other hand, American steel importers state that Japan loans its steel industry less than 1 percent of its new capital requirements, and that interest rates are at the regular bank rate.[15] In fact, government controlled banks, such as the Reconstruction Finance Bank, Japan Development Bank, Industrial Bank of Japan, and Long-Term Credit Bank, appear to have supplied the capital for well over half of the post-World War II growth in Japanese steel capacity.[16]

"Overpriced" Government Purchases Apart from the preferential margins given domestic products purchased by governments, special contracts with domestic suppliers may also contain an element of subsidization. Agreements between European public utilities and domestic suppliers commit governments to long-term purchasing at prices which may change at the option of the seller.[17] Such discretion permits clandestine subsidization, but we remain ignorant about the extent of such dealings, or whether they include iron and steel manufacturers.

Another example of this subsidization comes from France, where fifty-six "program contracts" have been negotiated between the government and steel industry, as well as some other manufacturers. Program contracts in France allow an industry to raise its prices in return for specific undertakings on productivity, salaries, investment programs, and exports.[18] In addition, France aids its steel industry in other unspecified ways.[19]

Internal Tax Rebates. In 1952 Japan inaugurated a system permitting steel firms to create additional reserves for bad debts as well as inventory and security revaluations—measures designed to lower tax liabilities.[20] Although accelerated depreciation and investment incentives also effect a decrease in taxes, the only general domestic rebate of an internal tax is found in the United Kingdom.

In 1966 the British government established a selective employment tax aimed at subsidizing the cost of labor in the manufacturing sector. By law, steel producers receive a full refund plus a rebate of 30 percent of their total tax liability.[21] The tax rate is $3.50 per week for adult male workers.[22] Congressional reports place employment in the British steel industry at 317,000.[23] Thus the annual tax for the British steel industry equals $57,694,000, but this is refunded in full. The rebate of 30 percent equals $17,308,200. Since wage payments represent variable costs, part of the rebate may be passed on to buyers of British steel. As a percentage of total revenue the rebate amounts to 0.6 percent.[24] If this figure is taken as the ad valorem reduction in steel costs, then the rate of subsidization is equal to 0.6 percent of British steel prices.

Input Subsidies. In effect, low-cost loans subsidize the input of capital, and employment tax rebates subsidize the factor of production called labor. But neither of those inputs receives the direct subsidization accorded to coal in the major steel-trading countries. To the extent that they are passed on, coal subsidies will lower the costs for domestic coal consumers. In turn, the lower cost could be shifted forward to the customers of coal users, resulting in a lower import demand for products competing with the coal users' output. The downward shift in import demand, as a percentage of the c.i.f. import price of the final output, constitutes the ad valorem rate of the nontariff barrier created by such a subsidy. We review the subsidies granted by each of the major steel-trading countries below.

In the Inner Six of the EEC, subsidies ranging from $.80 per ton in the Netherlands to $6.20 per ton in Belgium were granted in 1965.[25] A new subsidization scheme for coal in the ECSC calls for a set rate of $1.70 per ton or a varied rate which averages the same amount, but which can climb to a maximum of $2.20 per ton. The entire amount of the subsidy must be passed on to steelmakers, since the new prices are required to equal January 1, 1967 list prices less the subsidy. Canada has granted a bounty of 4.95 Canadian dollars per ton to producers of coke for domestic iron and steel manufacturers.[26]

The United Kingdom's National Coal Board sets British prices according to world prices, and has succeeded in raising British exports of coal to six million metric tons.[27] Such actions coupled with the board's failure to close inefficient mines have required large, direct government subsidization which amounted to $24 million in fiscal 1969.[28]

As part of its reconstruction program, Japan subsidized coal production. Since then, imported coal, which accounts for 84 percent of Japanese consumption has been freed from foreign exchange import quotas, and the government has authorized overseas investments in developing coal and iron ore reserves. However, no available information indicates the existence of per unit subsidies.[29]

Table 7-3 summarizes the rate of subsidization for coal in each of the major steel-trading countries in the first column. We derive the input subsidy per ton of steel output in Table 7-3 by multiplying the per unit coal

Table 7-3
Nontariff Barriers Created by Coking Coal Subsidies

	Coal Subsidy per Ton of Coal	*Coal Subsidy per Ton of Steel*	*Ad valorem Rate of Subsidization*
Belgium	$1.70	$1.12	1.3%
Canada	4.95	3.27	3.4
France	1.70	1.12	1.0
West Germany	1.70	1.12	1.2
Italy	1.70	1.12	0.8
Japan	0.00	0.00	0.0
Luxembourg	1.70	1.12	1.3
Netherlands	1.70	1.12	1.3
United Kingdom	NA	NA	NA
United States	0.00	0.00	0.0

Sources: See text for specific rates. Ad valorem rates are based on specific rates and import prices obtained from Statistical Office of the United Nations, *Monthly Bulletin of Statistics* (New York: United Nations, December 1968), pp. 192, 163.

subsidy by a coal-steel ratio of 0.66.[30] We can see the rate of subsidization as a percentage of c.i.f. import price in the third column. We convert the rate of subsidization to an ad valorem nontariff barrier rate in Chapter 4.

Indirect Tax Policy

As stated in Chapter 6 on foreign economic policies, the rationale of border-tax adjustment lies in the equalization of the internal taxes on imports and domestically produced goods. Following the same line of reasoning used to estimate the extent of subsidization in export rebates, this section considers only the amount by which border taxes exceed domestic taxes on the sale of the same product to be a quasitariff on imports.

Brief descriptions of internal, indirect taxation in each of the major steel-trading countries also appear in the discussion of tax rebates. Table 7-4 summarizes the range of internal and border taxes on iron and steel mill products. When the border tax exceeds the internal tax rate, the difference constitutes the ad valorem rate of this nontariff barrier.

Most of the major steel-trading countries do not tax imports at a higher rate than domestic goods. For instance, France and Germany tax imports

Table 7-4
Border Taxes on Imports[a]

	Internal Indirect Tax Range	*Border-Tax Range*	*Internal Indirect Steel Tax*	*Border Tax on Steel*	*Excess of Border over Internal Tax On Steel*
Belgium	1.00-16.00	11.35-16.85[b]	8.35	14.93	6.58
Canada	0.00-12.00	1.00-12.00	0.00[c]	12.00	12.00[c]
France	6.00-20.00	6.00-20.00	20.00	20.00	0.00
Germany	5.50-11.00	7.00- 9.00	11.00	7.00	0.00
Italy	0.60-30.00	7.20-37.80	4.00	8.80	4.80
Japan	0.00-40.00	0.00-40.00	0.00	0.00	0.00
Luxembourg	0.00-17.00	0.00- 7.00	3.09	3.09	0.00
Netherlands	1.72-31.25	1.72-40.65	5.25	11.00	5.75
United Kingdom	0.00-50.00	0.00-50.00	0.00	0.00	0.00
United States	0.00-15.00	0.00-15.00	4.80	4.80	0.00

[a]As a percentage of c.i.f., duty-paid, import price excluding the tax.

[b]For steel only.

[c]For non-construction steel only.

Sources: Internal tax range to internal tax on steel; see text.
Border-tax range and border tax on steel; see text.
Excess of border over internal tax is column 4 minus column 3.

at the same rate as their value-added tax on domestic products. Japan and the United Kingdom generally tax only consumer goods at the retail level and therefore do not usually apply border taxes. Luxembourg applies its domestic sales tax equally to imports, although it imposes a separate import tax discussed under import surcharges above. The United States generally levies taxes only at the retail level, and even with its mixed single-stage and cascade-tax systems at the state level, taxes are rarely levied at the border.[31] Four major steel traders, however, have imposed higher taxes on imports than on domestic steel.

Border Taxes on Steel. According to congressional reports, Belgium subjects steel imports not only to the normal turnover tax but also to an additional charge called majoration, the range of which varies considerably.[32] For steel imports, majoration varies from 3 percent to 8.5 percent with an unweighted average of 6.58 percent. The fact that there appears to be no relation between the stage of processing and the rate of majoration, seems to indicate that the extra tax is imposed for reasons other than compensating for tax occulte.[33]

The Canadian federal sales tax ranges from zero on most steel sales to 12 percent on sales of steel for construction purposes.[34] However, Canada taxes all steel imports at the 12 percent rate. To the extent that these imports are not for construction purposes, a higher-than-domestic tax is levied upon them.[35] The proportion of United States steel consumption accounted for by the construction sector approximated 20 percent in 1966.[36] Assuming that this ratio applies equally to Canadian steel consumption, one may conclude that the same proportion of Canadian imports would go to construction if distortions caused by trade restrictions were absent. Thus the excess border tax of 12 percent would apply to about 80 percent of Canadian steel imports.

Italy levies a so-called equalization tax of 3.6 to 7.8 percent on its steel imports, in addition to the 4 percent domestic turnover tax. The average equalization levy amounts to 4.8 percent.[37]

Restrictiveness. United States steel producers regard border taxes on imports as the most restrictive nontariff barrier imposed by Common Market countries. However, they make no distinction between the domestic-equivalent border taxes of France and Germany and the extra taxes added under the cascade-tax systems of other EEC member states. Indeed, United States producers regard all these border taxes as discriminatory, because from their point of view the higher European taxes penalize their exports more than their United States sales. They fail to appreciate the fact that their exports are taxed at the same rate as competing goods in the value-added countries.[38] United States producers find no border-tax barriers in the other major steel-trading countries, nor do United States importers find them in America.

Price Controls

Domestic price regulation may extend to imports and prevent them from competing pricewise with domestic goods, or in order to make imports less competitive they may lower domestic prices which are normally higher. These schemes for influencing aggregate or individual prices vary from the direct controls of the pure managed economy to indirect government pressure in what is otherwise a free enterprise system. While direct price support for some agricultural commodities is common practice among the industrial countries of the West, little direct control over prices of manufactures exists.

However, both the European Coal and Steel Community (ECSC) and individual European governments influence domestic steel prices. The ECSC encourages collusive pricing by making the publication of prices obligatory. In addition, the ECSC High Authority has the power to establish both maximum and minimum prices.[39] We have little evidence as to the extent this power has been exercised, but Dutch steelmakers have recently asked the Common Market Commission for a "relaxation of price control."[40]

On the eve of the ECSC, many European governments exerted some form of price control over steel producers. The post-World War II German government had a ceiling on steel prices until 1952, and afterwards permitted prices to be established by negotiation between producers and consumers. Maximum prices were also fixed by the French and Italian governments. In Belgium domestic steel prices were fixed by agreement between the steel producers and Fabrimetal, their main customer. Both here and in Luxembourg, however, the steel prices were subject to government approval. In contrast, the Netherlands provided for government price control only when it was necessary.[41]

Although the ECSC has not authorized price agreements among firms, internal prices still change simultaneously within each country. Apparently, such behavior results from the remaining government price control systems in each European country where "steel firms discuss prices among themselves and negotiate collectively with their respective governments through their national trade associations."[42]

Strikingly similar negotiations carry on between the JISF and the MITI.[43] The United States steel industry's pricing policies have been subjected to political pressure, but no direct price controls existed prior to August 15, 1971. Since then, of course, all American products continue to face on-again, off-again price controls.

Great Britain, the only major steel-trading country with a nationalized steel industry, is distinguished from other steel countries by its direct price controls. According to United States steel producers, these direct controls in the United Kingdom presented the only significant barrier of this type among the major steel traders, and it was ranked third in restrictiveness.

Patents

All the industrial countries of the West have laws protecting commercial property rights. Few appear to restrict importers more than domestic producers, although the foreign exporter may have more difficulty in learning about and satisfying domestic requirements from a distance.

Exporters do encounter difficulties and delays in obtaining simultaneous patent protection in many countries. However, the EEC and EFTA have been negotiating a centralization of procedures for a common European patent. Steel importers regard the United States patent system as restrictive to imports, although they give no specific indication of the type of obstacle it presents.

Other Internal Policies

Technical Specifications

The Common Market Commission once ranked stringent design and performance regulations as the most restrictive inter-European nontariff barrier, after taxes and quantitative restrictions. The EEC set the end of 1969 as the deadline for alignment of these standards. European government standards for textiles, motor vehicles, agricultural tractors and machinery, crystal glassware, electrical machinery and equipment, measuring instruments, and oil pipelines all have differed from one country to another. Oil pipeline provisions were of greatest relevance to primary steel producers. According to the EEC, varying methods used for calculating capacity and rules on the use and assembly of materials and equipment force pipe producers to custom-make pipe for each country "and thus lose the benefit of economies of scale."[44]

EFTA also took a first step toward aligning technical requirements by establishing a system of reciprocal use of test reports. According to an EFTA report, "the obligation to undergo separate tests for each individual market can greatly increase exporters' costs and thus constitute an indirect barrier to trade."[45]

In the United States examples of outright discrimination against foreign goods through technical regulations come from many states and localities. For instance, some state and local governments require boilers and containers for liquid gas under pressure to comply with the code of the American Society of Mechanical Engineers (ASME). State employees responsible for inspection must be affiliated with the ASME, "which however follows a policy of declining to approve manufacturers outside the United States and Canada . . ."[46]

United States steel importers have ranked technical regulations as the fifth most restrictive nontariff barrier in the United States. There have been no specific complaints made public, however, except those against changes in specifications which may discriminate against American imports. One such change has been proposed to the American Society of Testing and Materials. It recommends that structural steel imports be marked as to manufacturer as well as country of origin in raised letters. At last report the proposal had not been accepted by the society, but was still under consideration.[47]

National Security Restrictions

Although the practice of building up industrial complexes for military reasons is not uncommon, the United States provides the prime example of abusing national security to protect domestic producers. Ever since the 1954 Trade Agreements Extension Act, the United States president has had the authority to control imports for security reasons.[48] He can restrict imports by levying unlimited tariffs or quotas.[49] The Office of Civil and Defense Mobilization received many applications for import relief since passage of the national security amendment, partly because applications to the Tariff Commission under the escape clause provision have been so unsuccessful.[50] Nevertheless, few items have been accorded this type of protection, oil being the most important exception.

The United States steel industry continuously publicizes the threat to national security caused by United States steel imports. Their 1965 justification for import relief consisted of an estimate that in a limited war five years hence which cut off waterborne imports, total steel requirements would be 22 million tons larger than 1965 United States steel capacity.[51] Apart from the unwarranted assumptions that exports would continue at present levels, and that present capacity would not grow over the next five years, this estimate ignores two other important facts. First, United States steel imports have helped satisfy increased United States demand during the last two limited wars in Korea and Vietnam. Second, the domestic industry imports over one-third of its basic raw material, iron ore, and nearly all of such important alloy materials as chrome and manganese ore.[52] For all that, a Senate study concluded that "if the United States would rely more and more on importing steel, it would gamble with the national welfare and the national security. . . ."[53]

Notes

1. Massel, *Competition and Monopoly*, p. 65.

2. J. van B. Dresser et al., "Competitive Development of the Japanese Steel Industry," in J. Cohen, ed., *Pacific Partnership: United States-Japan Trade* (Lexington, Mass.: D. C. Heath, 1972), p. 203.

3. Senate Finance Committee, *Steel Imports*, p. 39.

4. *Time* (July 11, 1969), p. 74.

5. House Ways and Means Committee, *Foreign Trade and Tariff Proposals*, pt. 1, p. 304.

6. "World Steelmen Meet But Don't Quite Weld," *Business Week* (November 18, 1967), p. 46.

7. OECD *The Iron and Steel Industry in 1967 and Trends in 1968* (Paris: 1968), Table 38.

8. A. Fetherolf, "Erosion," *Steelways*, May 1961.

9. Joint Economic Committee, *Steel Prices*, p. 432; and J. van B. Dresser et al., "Competitive Development," p. 204.

10. British Iron and Steel Federation, "Rationalization in France," *Steel Review* (January 1967), p. 1.

11. J. Bell, "Fall of the House of Krupp," *Fortune* (August 1967), pp. 73, 74; BISF, "Rationalization in France," p. 1; Senate Finance Committee, *Steel Imports*, p. 40.

12. "Iron and Steel Industry," *The Oriental Economist*, April 1967.

13. AISI, *The Steel Import Problem*, New York, 1967, p. 30.

14. Senate Finance Committee, *Steel Imports*, p. 219.

15. *Newsletter*, West Coast Metal Importers Association, Inc., 2, no. 8 (August 1968), p. 3.

16. J. van B. Dresser et al., "Competitive Development," pp. 203-205.

17. House Ways and Means Committee, *Foreign Trade and Tariff Proposals*, pt. 8, p. 3663.

18. *TWA Marketair Newsletter*, "Spotlight on France," Transworld Airlines, 3, no. 5 (September 1967), p. 2.

19. *European Community* (August 1968), p. 5.

20. Dresser, et al., "Competitive Development," p. 204.

21. Senate Finance Committee, *Steel Imports*, p. 40.

22. *Time* (September 16, 1966), p. 112.

23. Senate Finance Committee, *Steel Imports*, p. 207.

24. Based on total revenue of $2,985.9 million. Source: Senate Finance Committee, *Steel Imports*, p. 455.

25. *Bulletin de la Communante,* Europeen du Charbon et de l'Acier Haute Autorite, no. 61. Luxembourg, (April 1966), p. 21.

26. Senate Finance Committee, *Steel Imports*, p. 34, 35, 39.

27. 1964 exports, see Kelly, "Nontariff Barriers," p. 272.

28. *Time* (July 11, 1969), p. 74.

29. Dresser et al., "Competitive Development," pp. 203, 213.

30. This output-input ratio is based on the 1322 pounds of coking coal consumed per ton of steel in the United States in 1966. Its use is qualified by the fact that subsidies and trade restrictions may have altered factor-use ratios, as well as by the fact that the ratios may differ among countries for other reasons. See Senate Finance Committee, *Steel Imports*, p. 169.

31. OECD, *Iron and Steel Industry*, pp. 116-175.

32. House Ways and Means Committee, *Foreign Trade and Tariff Proposals*, p. 3881.

33. Belgian Office of Foreign Trade, *Tariff des Droits Entree*, Ch. 73 and Annex, April 1969.

34. Letter dated June 23, 1969, United States Embasssy, Ottawa, Ontario.

35. Letter dated July 15, 1969, Department of National Revenue, Ottawa.

36. Senate Finance Committee, *Steel Imports*, p. 351.

37. Letter dated June 25, 1969, United States Embassy, Rome, Italy; and Senate Finance Committee *Steel Imports*, p. 37.

38. AISI, *The Steel Import Problem*, p. 32.

39. L. Lister, *Europe's Coal and Steel Community* (New York: Twentieth Century Fund, 1960), pp. 206, 210.

40. *European Community*, (May 1969), p. 23.

41. Lister, *Europe's Coal and Steel Community*, pp. 196, 197.

42. Ibid, p. 198.

43. Senate Finance Committee, *Steel Imports*, pp. 328-29.

44. *European Community* (May 1969), pp. 12, 13, 17.

45. *EFTA Reporter* (February 21, 1969), p. 3.

46. *European Community* (July 1965), pp. 9, 10.

47. West Coast Metal Importers Association, *Newsletter* (January 1969), p. 2.

48. Masson and Whitely, *Barriers to Trade,*, p. 41.

49. Subcommittee on Customs, Tariffs, and Reciprocal Trade Agreements, Committee on Ways and Means, *United States Customs, Tariff and Trade Agreement Laws and Their Administration* (Washington, D. C.: United States Government Printing Office, 1957), p. 100.

50. *The New York Journal of Commerce* (February 23, 1959), p. 1.

51. AISI, *The Steel Import Problem,* p. 50.

52. House Ways and Means Committee, *Foreign Trade and Tariff Proposals*, p. 2095.

53. Senate Finance Committee, *Steel Imports*, p. 246.

8 Restrictive Business Practices

We must include private practices in this study of governmental nontariff barriers because the governments of the Western industrial nations have long exercised control over restrictive business behavior. Thus their failure to take action against those practices which restrict imports implies that the governments sanction those restrictions. In this chapter a general discussion of restrictive business practices engaged in by steel cartels precedes a consideration of two specific practices which are claimed to deter United States steel exports: exclusive supply agreements coerced by steel cartels and freight rate discrimination practiced by ocean shipping cartels.

Steel Cartels

Since 1953, European steel producers have participated in an export cartel called the Entente de Bruxelles, whose avowed purpose was the fixing of minimum export prices for all the major steel products. Japan has export agreements among steel producers for particular steel products. The main purpose of these organizations has been to establish minimum export prices, set the total level of exports to the United States, and assign a certain share of the United States market to each of their members, who account for 85 percent of Japanese steel production.[1]

As for domestic cartels, there is no evidence of formal cartels in the Benelux countries or Italy, but Holland and Luxembourg have no need for them because of virtual monopolies in both countries. In Italy, a *de facto* steel cartel exists because the industry is dominated by the Finsider group which in turn has the government for a major stockholder. Besides marketing functions, four German cartels will attempt to guide members' investment and specialization policies, set production quotas, and maintain a uniform price system. French steel producers make joint decisions on matters of supply, marketing, domestic and foreign sales, investment, and raw materials.[2] While the oligopolistic United States steel industry would otherwise be receptive to cartelization, American antitrust laws prevent overt agreements. Despite this, the United States industry has tried to avoid the characteristic price instability of oligopoly by occasionally accepting its dominant firm as a price leader.[3]

Restrictive Practices

As evidenced by the goals of various cartels, they participate in a wide variety of restrictive business practices. They fix prices at high levels in order to make excess profits, or may lower them below cost to drive out competition. If they can separate the markets in which they sell, cartels may practice price discrimination, charging each set of buyers different prices, according to what each market will bear. They allocate product or geographical markets or shares of markets among their members so as to prevent any need for rivalry. They may attempt to maintain their control of certain markets through exclusive supply agreements, and monopolization of patents or other factors of production.

Monopolies practice many of the same activities, but there is a constraint on cartel activity that does not apply to monopolies: cartel policies will be successful only insofar as its members are willing to follow them. In the United States, where agreements in restraint of trade are illegal, cartels find it difficult to enforce their rules.[4] Despite government sanctions, European cartels also face problems dealing with intransigent members. The original international steel cartel's control schemes were abandoned by its members at the start of the depression.[5] Even in national steel cartels, price-cutting in violation of the agreement occurs during recession.[6] However, the French have a new penalty for divergence from quotas and prices. Each steel producer now deposits a blank check made out to the cartel. If the producer breaks the agreements, his fine is written in and the check is cashed, proceeds going to the other members.[7]

Another constraint applies to both monopolies and cartels: their policies will be successful only so long as the cartels are immune from outside competition. National agreements may not include all producers, and excess profits may attract new entrants to an industry. New entrants may come from domestic sources or from abroad. If they cannot be driven out through price wars or exclusive supply agreements, cartels may seek to exclude them by political means. In the case of competition from abroad, cartel pressure may cause governments to erect tariffs and other barriers to imports. Thus, apart from any nontariff barriers established by the cartel itself, restrictive agreements may also promote governmental barriers. Of course, for international cartels the problems of outside competition are magnified, and the resort to political protection is not feasible in the markets of nonmembers.

To what extent do cartels alone raise obstacles to imports? Charging a higher price in the domestic market and a lower one abroad creates no import barriers between the exporting and importing countries, and most industrial countries have antidumping restrictions which restrict the effectiveness of such practices. In third markets, however, the lower prices charged by cartels may effectively exclude nonmember exporters. Predatory price-cutting by a cartel or a monopolist can raise an important barrier to the entrance of imports. Exclusive supply agreements can also place imports at a competitive disadvantage, and this is discussed in the next section. As for the remaining restrictive business practices, however, we can say that in themselves they do more to attract imports by raising the domestic price above a competitive level.

Theoretically, the nontariff barriers created by foreign steel cartels and monopolies are few, but there is no way to quantify their effects on imports. World steel producers consider cartel activities in general to be significant barriers in all major steel countries except Canada. To them, the British Steel Corporation is little different from the government-controlled cartels of Europe, and they consider its policies to be the second most important barrier in the United Kingdom. Import restrictions by French and German cartels rank third in each country. Those imposed by Italian and Benelux ententes rank fourth among their import barriers, and those of Japan are in fifth place.

Exclusive Supply Agreements

Although quality differences fail to make steel a differentiated product because such differences are measurable,[8] it is possible for some steel producers to monopolize the output of certain types or qualities of steel. Moreover, large-volume buyers with variable production levels may be constrained to buy from nearby producers to assure continuous supply, quick delivery, and reduce the necessity to carry large inventories. Whenever domestic producers obtain a degree of monopoly power because of these conditions, they may exercise it through exclusive supply agreements. These are similar to tie-in contracts by which the customer agrees to purchase all required steel from the domestic producer in order to be able to buy any.[9]

American steel producers claim that United States exports are restrained by steel cartels such as the "Tin Plate Club" in Europe. The main restrictive practice employed by such cartels is a threat to cut off a customer's main source of supply if it buys imported steel. In this study's survey of United States steel exporters, exclusive supply agreements were ranked as the fifth most restrictive barrier in Germany, Italy, the Benelux countries, and Canada, while they were ranked sixth in France, and found to be insignificant in Japan and the United Kingdom. United States steel importers claim that these agreements are the fourth most important barrier to American imports.

Freight Rate Discrimination

The ocean freight rates applicable to foreign trade among the major steel countries are for the most part established by steamship cartels known as *conferences*. The steamship conferences were formed to eliminate price competition among the member lines, to standardize shipping practices, and to provide regularly scheduled service between designated ports.[10]

Each conference publishes ocean freight-rate schedules or tariffs on a commodity basis, with different rates applied to varying commodities for movement between the same points. Each conference publishes hundreds and sometimes thousands of rates, and whenever no specific rate applies, the steamship line assigns a general cargo rate. Carriers quote rates on the basis of weight, space occupied, or a combination of both, but they usually levy steel rates on a per ton basis.[11]

The United States Shipping Act requires steamship lines serving the United States foreign trade as common carriers to file their tariffs with the Federal Maritime Commission, maintain them in an up-to-date manner, and keep them open to public inspection.[12] The law also requires that only those rates on file with the commission can be charged, and it may disapprove any rate which, after hearing, it finds to be "so unreasonably high or low as to be detrimental to the commerce of the United States." Moreover, the Shipping Act prohibits rates which are unjustly discriminatory between shippers or ports, or unjustly prejudicial to exporters of the United States as compared with their foreign competitors. When a given ocean freight rate restricts his exports, a United States exporter must first enter into negotiation with the particular conference involved, and then make an informal or formal complaint to the Federal Maritime Commission.

The lack of price competition among shipping lines is the most frequently mentioned disadvantage of the conference system for exporters. Critics of the system claim that the level of ocean freight rates climbs higher than it would be if the forces of competition were freely at play. Exclusive patronage arrangements, such as the dual rate contract, serve to reduce the choice available to exporters when services are available from independent carriers. The dual rate contract consists of an arrangement whereby an exporter, in exchange for committing all or a fixed portion of his shipments to the vessels of a given conference, receives a "contract rate" that may be as much as 15 percent below the published tariff rate that applies to exporters who do not sign exclusive patronage contracts.

Contract rates assume the most significance for purposes of comparison, since the availability of independent carriers appears to be limited. In 1961 independent liners carried only 5 percent of United States exports and 10 percent of United States imports.[13] Reportedly, an equally small percentage of steel imports entered the United States on tramp steamers.[14]

Nevertheless, analysis of these rates must be qualified by the fact that independents are handling an increasing share of steel trade volume. Moreover, major steel shippers often seek to negotiate lower rates than those listed.[15] Although freight rate differentials on exports to third countries also exist, these will not be considered here.

Senator Paul Douglas first publicized alleged freight rate discrimination against American exports.[16] He claimed that export rates were generally 30 to 40 percent higher than import rates, due to European and Japanese control of international shipping cartels.[17] Paradoxically, the United States steel industry has never listed freight rate discrimination among the many nontariff barriers it claims to face. Five of seven major steel companies surveyed by the Council of Economic Advisors replied that "ocean shipping rate differentials do not put them at a major disadvantage" in export markets.[18]

Reacting to Senator Douglas's findings, one specialist in steel freight rates commented that "published conference rates . . . mean very little unless there is an active trade in the product involved."[19] According to industry spokesmen, the rates compare favorably on those steel products which account for the largest share of United States exports.[20] However, it is possible that the causal relationship runs in an opposite direction from that implied by the industry spokesmen. The present composition of United States exports could stem from low export freight rates on some items, and American firms may not sell other products abroad because of high rates.

Tables 8-1 and 8-2 present the actual rates in question between the United States and other overseas steel traders. The lower contract rates are presented since these apply to most large shippers. Similar comparisons have been made for 1962, 1966, and 1967, but they do not cover the entire range of steel products, nor do they show the rates for all relevant trade routes.[21] For these reasons, we use 1963 rates. Inspection of available 1967 data indicates that although freight rates rose, differentials remained approximately the same. We obtained rates for all SITC steel products except pig iron and hoop and strip.[22] Although differences exist between the statistical classification of commodity trade on the one hand, and freight rate tariffs on the other, disaggregation of the statistical data to the three- and four-digit level permitted a close matching.

Senator Douglas made his estimate of the freight rate differential on the basis of an unweighted average of the shipping cost per ton. Shipping conference representatives expressed their estimates on the average differential per pound of steel weighted by the United States trade volume of each particular item.[23] Another contribution considered the freight rate differentials according to the volume of each item in United States foreign trade.[24] With regard to a small sample of United States imports in SITC categories 5-8, the latter study found a positive differential (i.e., export minus import rate as a percentage of c.i.f. value) of 2.5 percent, while United States manufactured exports have a positive differential of 0.7 percent.[25]

The same study found that transportation charges varied from 3 to 5 percent of the landed value of manufactures. In contrast, steel freight rates range from 4.6 to 27.2 percent of the c.i.f. steel price. The export rates shown in Table 8-3 have an unweighted average of 10.7 percent, while import rates in Table 8-4 average 12.8 percent. This negative differential in terms of United States steel trade reflects the fact that some export rates per ton are lower than import rates, as well as the disparity between high United States steel export prices and low import prices. The differentials by steel product and by country are presented in Table 8-5.

As noted above, weighting the freight rate differentials by the share of each steel product in United States steel trade can result in the protective effect of the differentials being understated. This understatement results from high differentials which may severely restrict certain products being given very little weight. Ideally, each differential should be weighted by what trade would have been in their absence. Since this is not known, a second-best weight might be the share of each steel product in the domestic steel consumption of each country. Although this information is readily available for the United States, it is not complete for Europe.[26] Therefore, this study has used the share of each steel product in the total export trade of all the main steel traders as weights for the freight rate differentials. We find their percentage shares listed in Table 8-6. The weighted differentials, with their sum equalling the average for each country vis-à-vis the United States, appear in Table 8-7. Negative differentials which restrict United States steel imports range from a high of 5.4 percent on trade with Italy to a low of 2.1 percent on trade with France.

The only positive differential appears in trade with Japan. Our survey of American steel producers, who cite Japan as the only importer with higher freight rates, supports that finding. The barrier created by this discrimination ranked last in importance, however. For some reason, United States importers did not list inbound freight rates as restrictive to United States steel imports.

To determine the average differential facing United States steel imports from all eight countries, we can weight each country's average differential by that country's share in the combined steel export trade of the group. Table 8-8 shows these shares for 1963. The weighted average U.S. freight rate differential is 3.26 percent of c.i.f. value. If the burden of adjustment is placed on inbound United States and Japanese rates, they must decrease by the amount of the differential in order to eliminate this nontariff import barrier. Therefore the differentials represent the ad valorem rate of this import barrier, and their restrictive effect on imports can be estimated like that of other quantified restrictions.[27]

Table 8-1
Conference Ocean Freight Rates on United States Exports to Major Steel-trading Countries (Dollars per Long Ton)

Steel Products	*Country of Destination*					
	Benelux[a]	*France*[b]	*Germany*[c]	*Italy*[d]	*Japan*[e]	*United Kingdom*[f]
Structural sections	25.75	28.50	28.50	27.50	28.10	25.75
Castings & forgings	36.25	40.25	40.00	46.50	55.50	36.25
Billets & blooms	13.25[g]	13.25	13.25	13.25	30.35	13.25
Rails	30.25	33.50	33.25	30.25	36.35	30.25
Wire rods	21.00	26.75	23.00	21.00	30.35	21.00
Pipes	46.25	51.25	51.00	41.25	30.35	46.25
Bars	13.25[g]	27.50	13.25	13.25	28.10	13.25
Sheets	13.25	13.25	13.25	20.00	30.35	13.25
Plates	15.25	13.25	15.25	20.00	24.10	15.25
Wire excl. rods	24.75	25.00	27.25	24.75	30.35	24.75

[a]Rates between Antwerp, Rotterdam, Amsterdam: and United States North Atlantic ports.

[b]Rates between French North Atlantic ports and United States Gulf ports.

[c]Rates between West Germany and United States North Atlantic ports.

[d]Rates between Italy and United States North Atlantic ports.

[e]Rates between Japan and United States Pacific Coast ports.

[f]Rates between Benelux and United States North Atlantic ports in the absence of data on the United Kingdom.

[g]Minimum rates.

Note: Canada is not reported to have discriminatory freight rates; shipping charges are also less significant between Canada and the United States.

Sources: Joint Economic Committee, *Steel Prices*, pp. 555-57, 563-66; Joint Economic Committee, *Discriminatory Ocean Freight Rates*, pp. 785, 805, 960-964.

Table 8-2
Conference Ocean Freight Rates on United States Imports from Major Steel-trading Countries (Dollars per Long Ton)

Steel Products	*Origin*					
	Benelux	*France*	*Germany*	*Italy*	*Japan*	*United Kingdom*
Structural sections	17.75	13.50	17.75	23.25	15.50	17.75
Castings & forgings	26.25	34.00	26.25	30.00	24.00	26.25
Billets & blooms	17.25	13.50	17.25	17.25	15.50	17.25
Rails	17.75	13.50	17.75	17.75	15.50	17.75
Wire rod	16.50	13.50	16.50	16.50	15.50	16.50
Pipe	18.75	14.50	18.75	24.75	18.00	18.75
Bars	17.75	31.00	17.75	17.75	15.50	17.75
Sheets	20.00	13.50	20.00	23.25	15.50	20.00
Plates	20.00	13.50	20.00	23.25	15.50	20.00
Wire excl. rods	15.75	15.00	15.75	15.75	22.75	15.75

Sources: Same as Table 8-1.

Table 8-3
Conference Ocean Freight Rates[a] on United States Exports to Major Steel-trading Countries (As a Percentage of United States Export Price[b])

Steel Products	*Destination*					
	Benelux	*France*	*Germany*	*Italy*	*Japan*	*United Kingdom*
Structural sections	14.0	14.0	14.0	14.0	27.2	14.0
Castings & forgings	8.5	10.6	9.2	8.5	11.8	8.5
Billets & blooms	4.6	9.2	4.6	4.6	9.4	4.6
Rails	15.1	16.4	16.4	15.0	16.2	15.1
Wire rods	6.5	5.7	6.5	8.4	9.9	6.5
Pipes	5.2	5.2	5.2	7.7	11.2	5.2
Bars	16.6	18.0	17.0	16.6	19.3	16.6
Sheets	8.0	8.1	8.8	8.0	9.7	8.0
Plates	9.3	10.3	10.2	8.4	6.3	9.3
Wire excl. rods	9.0	9.9	9.9	11.3	13.2	9.0

[a]Based on freight rates on Table 8-1.

[b]C.i.f. prices calculated from data on f.o.b. prices and freight charges in Joint Economic Committee, *Discriminatory Ocean Freight Rates,* p. 12; and United Nations, *Statistical Papers,* "Commodity Trade Statistics," 1963, Series D, Volume XIII, Numbers 1-20, pp. 4042, 4809.

Table 8-4
Conference Ocean Freight Rates[a] on United States Imports from Major Steel-trading Countries (As a Percent of United States Import Price[b])

Steel Products	*Origin*					
	Benelux	*France*	*Germany*	*Italy*	*Japan*	*United Kingdom*
Structural sections	15.3	12.1	15.3	19.2	13.6	15.3
Castings & forgings	6.9	8.8	6.9	7.8	6.3	6.9
Billets & blooms	18.1	14.7	18.1	18.1	16.6	18.1
Rails	15.7	12.4	15.7	15.7	14.0	15.7
Wire rods	14.7	12.3	14.7	14.7	13.9	14.7
Pipes	11.2	8.9	11.2	14.3	10.8	11.2
Bars	11.9	19.0	11.9	11.9	10.5	11.9
Sheets	10.0	7.0	10.0	11.4	7.0	10.0
Plates	17.7	12.7	17.7	20.0	14.3	17.7
Wire excl. rods	8.8	8.4	8.8	8.8	12.2	8.8

[a]Based on freight rates in Table 8-2.

[b]C.i.f. prices calculated from data on f.o.b. prices and freight charges in same sources as in Table 8-3.

Table 8-5
Conference Ocean Freight Rate[a] Differentials[b] (As a Percentage of United States C.I.F. Price)

Steel Products	*Between the United States and:*					
	Benelux	*France*	*Germany*	*Italy*	*Japan*	*United Kingdom*
Structural sections	− 4.1	−0.7	− 4.1	− 4.1	10.6	− 4.1
Castings & forgings	− 6.2	−1.7	− 5.5	− 6.2	− 2.9	− 6.2
Billets & blooms	− 7.3	−9.8	− 8.3	− 7.3	− 1.1	− 7.3
Rails	− 0.2	4.3	1.1	− 3.3	2.6	− 0.2
Wire rods	−11.2	−7.0	−11.2	−11.6	− 4.4	−11.2
Pipes	− 4.8	−1.8	− 4.8	− 3.7	3.3	− 4.8
Bars	0.9	5.6	2.2	0.9	5.3	0.9
Sheets	− 0.8	−0.3	0.0	− 0.8	− 2.5	− 0.8
Plates	− 1.9	1.4	− 1.0	− 5.9	− 4.5	− 1.9
Wire excl. rods	2.1	1.1		3.5	6.9	2.1

[a]Rates obtained from Tables 8-3 and 8-4.

[b]United States export rates minus import rates.

Table 8-6
The Share of Steel Products in the Total Steel Trade of the Major Steel-trading Countries[a]

Steel Products			
SITC Code	*Description*	*Share of Total*[b]	*Percentage of Total*
673.4,5	Structural sections	458,668	9.9
679.0	Castings & forgings	18,360	0.4
672.0	Billets & blooms	314,083	6.8
676.0	Rails	93,030	2.0
673.1	Wire rods	216,804	4.7
678.0	Pipes	900,697	19.5
673.2	Bars	600,152	13.0
674.3,7,8	Sheets	1,420,418	30.8
674.1,2	Plates	381,151	8.3
677.0	Wire excl. rods	211,539	4.6
Total		4,614,902	100.0

[a]Total exports of Benelux, France, Germany, Italy, Japan, United Kingdom, and the United States. Canada is excluded.

[b]F.o.b. value in thousands of United States dollars (1963).

Source: United Nations, *Statistical Papers,* "Commodity Trade Statistics," 1963, Series D, Vol. XIII, No. 1-20, pp. 4039-4042; No. 1-23, pp. 4574-4578; No. 1-19, pp. 3815-3818; No. 1-18, pp. 3566-3570; No. 1-22, pp. 4392-4395; No. 1-20, pp. 3871-3873; No. 1-12, pp. 2247-2251; No. 1-15, pp. 2864-2869.

Table 8-7
Conference Ocean Freight Rate Differentials[a] Weighted by the Share of Each Steel Product in Total Steel Exports[b]

Steel Products	*Between the United States and:*					
	Benelux	*France*	*Germany*	*Italy*	*Japan*	*United Kingdom*
Structural sections	− .279	− .476	− .279	− .279	.721	− .279
Castings & forgings	− .291	− .130	− .258	− .291	− .136	− .291
Billets & blooms	− .949	−1.274	− .949	− .949	− .143	− .949
Rails	− .020	.693	.109	− .327	.251	− .020
Wire rods	− .930	− .581	− .930	− .963	− .365	− .930
Pipes	−1.478	− .554	−1.478	−1.140	1.016	−1.478
Bars	.018	.112	.044	.018	.106	.018
Sheets	− .368	− .138	.000	− .368	− .115	− .368
Plates	− .370	.273	− .195	−1.150	− .877	− .370
Wire excl. rods	.008	.004	.012	.014	.028	.084
Total[c]	−4.651%	−2.071%	−3.924%	−5.435%	.492%	−4.651%

[a]As a percentage of c.i.f. price from Table 8-5.

[b]From Table 8-6.

[c]Column totals are equal to the average differential for all steel products in trade between the United States and each of the above countries.

Table 8-8
United States Freight Rate Differential[a] Calculation

Other Major Steel Traders	*Average Freight Rate Differential*[b]	*Share of Each Country in Total Steel Trade*[c]	*Weighted Freight Differentials*[d]
Benelux	4.65	26.0	1.21
France	2.07	16.0	0.33
Germany	3.92	25.0	0.98
Italy	5.43	4.0	0.22
Japan	−0.49	16.0	−0.08
United Kingdom	4.65	13.0	0.60
Average United States freight rate differential			3.26

[a]United States import minus export rate as a percent of c.i.f. price. This is equal to the sum of the third column.

[b]Based on differentials from Table 8-7 but with opposite sign.

[c]Calculated from 1963 f.o.b. value of exports. Source: United Nations (See note [h], Table 8-3). Share is shown as a percentage of total steel exports from overseas steel traders.

[d]Percentage of c.i.f. price. Derived by multiplying the first column times the second. Sum of these weighted differentials gives the average United States freight rate differential.

Notes

1. Senate Finance Committee, *Steel Imports*, pp. 328-29.

2. "World Steelmen Meet But Don't Quite Weld," *Business Week* (November 18, 1967), p. 98.

3. A. Kaplan, J. Dirlam, R. Lanzillotti, *Pricing in Big Business* (Washington, D. C.: The Brookings Institution, 1958), pp. 13-23.

4. For part of the extensive literature on the extent of cartel activity under the antitrust laws, see D. Orr and P. W. MacAvoy, "Price Strategies to Promote Cartel Stability," *Economica* (May 1965), pp. 186-197; G. Stigler, "A Theory of Oligopoly," *Journal of Political Economy* (February 1964), pp. 44-61.

5. Heuser, *Control of International Trade*, p. 79.

6. Lister, *Europe's Coal and Steel Community*, pp. 205-206.

7. "World Steelmen Meet," *Business Week*.

8. R. Caves, *American Industry: Structure, Conduct, and Performance* (Englewood Cliffs, N. J.: Prentice Hall, 1967), p. 21.

9. GATT, *Restrictive Business Practices*, 1959, p. 25.

10. United States Department of Commerce, Federal Maritime Commission, *Ocean Freight Rate Guidelines for Shippers*, Washington, D. C.: United States Government Printing Office, May 1966, p. 1. .

11. Senate Finance Committee, *Steel Imports*, p. 415.

12. The following two paragraphs draw on *Ocean Freight Rate Guidelines*, pp. 3, 7, 10.

13. Joint Economic Committee, *Steel Prices*, p. 551.

14. House Ways and Means Committee, *Foreign Trade and Tariff Proposals*, pt. 4, p. 1702.

15. Senate Finance Committee, *Steel Imports* p. 414.

16. Joint Economic Committee, *Steel Prices*, pp. 530-70.

17. *Hearings on Senate Resolution 149*, p. 87.

18. Council of Economic Advisers, *Report to the President on Steel Prices*, 1965, p. 17.

19. "Congressional Confusion: Steel 'Inequities' Hit by Douglas," *American Metal Market* (May 3, 1963), p. 1.

20. Ibid.

21. Joint Economic Committee, *Discriminatory Ocean Freight Rates and the Balance of Payments*, 88th Cong., 1st sess., (Washington, D. C.: United States Government Printing Office, 1963), pt. 1, p. 10; Senate Finance Committee, *Steel Imports*, p. 415-17, and Thorn, "Steel Imports," p. 379.

22. Standard Industrial Trade Classification Numbers 671 and 675.

23. Joint Economic Committee, *Discriminatory Ocean Rates*, pt. 5, pp. 777-987.

24. Joint Economic Committee, 89th Cong., 1st sess., *Discriminatory Ocean Freight Rates and the Balance of Payments* (Washington, D. C.: United States Government Printing Office, pt. 3), p. 450.

25. Ibid, p. 471.

26. OECD, *The Iron and Steel Industry in 1967 and Trends in 1968*, Table 26.

27. This analysis ignores freight rate differentials on steel inputs like coal and iron ore. Cf. Thorn, "Steel Imports," p. 378.

Appendix: Statistical Tables

Table A-1
United States Imports of Iron and Steel Mill Products

Year	*Quarter*	*Quantity*		*Value*		
		1,000s of Short Tons	*Index 1957-59 = 100*	*Total in $1,000s*	*Unit per Ton in Dollars*	*Index of Unit Value 1957-59 = 100*
1954	I	207.0	31.96	22,700	109.66	78.86
1954	II	324.0	50.02	30,900	95.37	68.58
1954	III	308.0	47.55	31,800	103.25	74.25
1954	IV	254.0	39.21	29,500	116.14	83.52
1955	I	253.0	39.20	29,500	116.19	83.55
1955	II	319.6	49.34	32,500	101.69	73.13
1955	III	364.2	56.23	36,500	100.22	72.07
1955	IV	427.0	66.06	46,500	108.67	78.15
1956	I	342.7	52.91	44,600	130.14	93.58
1956	II	427.0	65.92	50,700	118.73	85.38
1956	III	455.7	70.35	56,900	124.86	89.79
1956	IV	591.7	91.35	77,700	130.47	93.82
1957	I	433.4	66.91	65,800	151.82	109.17
1957	II	367.8	56.78	60,100	163.40	117.50
1957	III	259.7	40.09	44,900	172.89	124.33
1957	IV	249.6	38.53	41,800	167.47	120.43
1958	I	269.6	41.62	41,000	152.08	109.36
1958	II	386.8	59.72	51,300	132.63	95.38
1958	III	536.5	82.83	63,000	117.43	84.44
1958	IV	644.3	99.47	75,200	116.71	83.93
1959	I	791.8	122.24	92,500	116.82	84.01
1959	II	1205.3	186.08	136,800	113.50	81.62
1959	III	1191.5	183.95	148,500	124.63	89.62
1959	IV	1436.5	221.77	200,100	139.30	100.17
1960	I	1470.0	226.95	206,400	140.41	100.97
1960	II	874.5	135.01	126,600	144.77	104.11
1960	III	618.2	95.44	87,700	141.86	102.01
1960	IV	607.4	93.77	84,800	139.61	100.39
1961	I	461.2	71.20	61,300	132.91	95.58
1961	II	714.0	110.23	84,600	118.49	85.21
1961	III	805.1	124.29	95,500	118.62	85.30
1961	IV	917.0	141.57	105,300	114.83	82.57
1962	I	895.6	138.27	106,700	119.14	85.67
1962	II	1046.2	161.52	122,300	116.90	84.06
1962	III	994.4	153.52	115,200	115.85	83.31
1962	IV	992.6	153.24	112,700	113.54	81.65

Table A-1 continued

		1,000s of Short Tons	*Index 1957-59 $ 100*	*Total in $1,000s*	*Unit per Ton in Dollars*	*Index of Unit Value 1957-59 $ 100*
1963	I	888.3	137.14	110,200	124.06	89.21
1963	II	1312.5	202.63	142,700	116.34	83.66
1963	III	1531.8	236.49	175,000	114.24	82.15
1963	IV	1403.0	216.60	163,000	116.18	83.55
1964	I	1304.0	201.32	154,000	118.10	84.93
1964	II	1565.0	241.61	181,000	115.65	83.16
1964	III	1522.0	234.97	178,000	116.95	84.10
1964	IV	1735.0	267.86	202,000	116.43	83.73
1965	I	1747.0	269.71	209,000	119.63	86.03
1965	II	3015.0	465.47	341,000	113.10	81.33
1965	III	2868.0	442.78	315,000	109.83	78.98
1965	IV	2463.0	380.25	275,000	111.65	81.23
1966	I	1851.0	285.77	219,000	118.31	85.08
1966	II	2559.0	395.07	289,000	112.93	81.21
1966	III	3207.0	495.11	356,000	111.01	79.83
1966	IV	2824.0	435.98	319,000	112.96	81.23
1967	I	2367.0	365.43	283,000	119.56	85.98
1967	II	2783.0	429.65	324,000	116.42	83.72
1967	III	2869.0	442.93	320,000	111.54	80.21
1967	IV	3285.0	507.15	363,000	110.50	79.46
1968	I	3340.0	515.65	383,000	114.67	82.46
1968	II	4709.0	727.00	515,000	109.36	78.64
1968	III	5284.0	815.77	572,000	108.25	77.84
1968	IV	4408.0	680.53	492,000	111.61	80.26
1969	I	1872.0	288.9	245,000	130.88	94.12
1969	II	4595.0	709.1	548,000	119.26	85.76
1969	III	3919.0	604.8	491,000	125.29	90.10
1969	IV	3377.0	521.1	440,000	130.29	93.69
1970	I	2275.0	351.1	352,000	154.73	111.27
1970	II	3058.0	471.9	464,000	151.73	109.11
1970	III	3484.0	537.7	518,000	148.68	106.92
1970	IV	4361.0	673.0	620,000	142.17	102.24
1971	I	3692.0	569.8	544,000	147.35	105.96
1971	II	5199.0	802.3	734,000	141.18	101.52
1971	III	4971.0	767.1	712,000	143.23	103.00
1971	IV	4200.0	648.1	624,000	148.57	106.84
1972	I	3206.0	494.8	509,000	158.76	114.17
1972	II	4051.0	625.2	642,000	158.47	113.96
1972	III	4805.0	741.5	756,000	157.34	113.15
1972	IV	5253.0	808.2	836,000	159.14	114.39

Sources: United States Department of Commerce, *World Trade Information Service* and *Overseas Business Reports*, Washington, D.C.: United States Government Printing Office, various years.

Table A-2
Steel Price Indexes (1957-59 Average = 100)

Year	*Quarter*	*Import*	*Domestic*	*Relative*
1954	I	78.9	77.5	101.8
1954	II	68.6	77.4	88.6
1954	III	74.2	78.9	94.1
1954	IV	83.5	79.0	105.7
1955	I	83.6	79.0	105.7
1955	II	73.1	79.1	92.4
1955	III	72.1	83.6	86.2
1955	IV	78.2	84.3	92.7
1956	I	93.6	85.5	109.4
1956	II	85.4	96.3	98.9
1956	III	89.8	90.1	99.6
1956	IV	93.8	92.4	101.6
1957	I	109.2	94.2	115.8
1957	II	117.5	95.0	123.7
1957	III	124.3	98.8	125.9
1957	IV	120.4	99.1	121.4
1958	I	109.4	99.4	110.0
1958	II	95.4	99.3	96.0
1958	III	84.4	100.7	83.8
1958	IV	84.0	101.8	82.5
1959	I	84.0	101.9	82.4
1959	II	81.6	102.0	80.0
1959	III	89.6	102.2	87.7
1959	IV	100.2	102.1	97.7
1960	I	100.9	102.0	98.9
1960	II	104.1	102.2	101.8
1960	III	102.0	102.0	100.0
1960	IV	100.4	102.0	99.4
1961	I	95.6	102.2	93.5
1961	II	85.2	102.0	83.5
1961	III	85.3	101.9	83.7
1961	IV	82.6	101.9	81.0
1962	I	85.7	101.8	84.1
1962	II	84.0	101.9	82.5
1962	III	83.3	101.8	81.8
1962	IV	81.7	101.8	80.2
1963	I	89.2	101.7	87.7
1963	II	83.7	102.1	81.9
1963	III	82.1	102.3	80.3
1963	IV	83.6	103.3	80.9

Table A-2 contintued

Year	*Quarter*	*Import*	*Domestic*	*Relative*
1964	I	84.9	103.4	82.2
1964	II	83.1	103.3	80.5
1964	III	84.1	103.1	81.6
1964	IV	83.7	103.0	81.3
1965	I	86.0	103.4	83.2
1965	II	81.3	103.6	78.5
1965	III	78.9	104.0	85.9
1965	IV	80.3	104.0	77.2
1966	I	85.0	104.7	81.2
1966	II	81.2	104.9	77.4
1966	III	79.8	105.4	75.7
1966	IV	81.2	106.0	76.6
1967	I	85.9	106.6	80.6
1967	II	83.7	106.6	78.5
1967	III	80.2	107.2	74.8
1967	IV	79.4	107.9	73.6
1968	I	82.5	108.9	76.1
1968	II	78.6	109.2	72.0
1968	III	77.8	110.2	70.6
1968	IV	80.2	111.0	73.3
1969	I	94.1	112.3	83.8
1969	II	85.8	113.7	75.5
1969	III	90.1	116.0	77.7
1969	IV	93.7	117.3	79.9
1970	I	111.3	118.3	94.1
1970	II	109.1	121.2	90.0
1970	III	106.9	123.5	86.6
1970	IV	102.2	124.7	82.0
1971	I	106.0	125.4	84.5
1971	II	101.5	128.4	79.0
1971	III	103.0	134.3	76.7
1971	IV	106.8	136.0	78.5
1972	I	114.2	138.0	82.8
1972	II	114.0	138.0	82.6
1972	III	113.2	138.4	81.8
1972	IV	NA	138.7	NA

Sources: *Wholesale Price Index* and United States Department of Commerce, *World Trade Information Service* and *Overseas Business Reports*.

Table A-3
Index of United States Gross National Product in Constant Dollars (1957-59 Average= 100)

Year	*Quarter*	*Index*	*Year*	*Quarter*	*Index*
1954	I	87.8	1964	I	124.5
1954	II	87.7	1964	II	126.2
1954	III	88.8	1964	III	127.7
1954	IV	90.6	1964	IV	128.3
1955	I	93.3	1965	I	131.2
1955	II	94.9	1965	II	133.1
1955	III	96.4	1965	III	136.7
1955	IV	97.3	1963	IV	138.6
1956	I	96.7	1966	I	141.4
1956	II	97.2	1966	II	142.4
1956	III	96.9	1966	III	143.8
1956	IV	98.2	1966	IV	145.5
1957	I	98.9	1967	I	145.1
1957	II	98.8	1967	II	145.9
1957	III	99.2	1967	III	147.3
1957	IV	97.7	1967	IV	148.7
1958	I	95.4	1968	I	151.0
1958	II	95.8	1968	II	153.4
1958	III	98.3	1968	III	155.3
1958	IV	100.6	1968	IV	156.6
1959	I	102.2	1969	I	157.5
1959	II	104.7	1969	II	158.3
1959	III	103.6	1969	III	159.0
1959	IV	104.7	1969	IV	158.1
1960	I	106.9	1970	I	157.3
1960	II	106.8	1970	II	157.5
1960	III	106.2	1970	III	158.6
1960	IV	105.5	1970	IV	156.9
1961	I	105.2	1971	I	160.3
1961	II	107.4	1971	II	161.5
1961	III	109.3	1971	III	162.9
1961	IV	111.6	1971	IV	165.5
1962	I	113.3	1972	I	167.5
1962	II	115.1	1972	II	171.3
1962	III	116.3	1972	III	173.7
1962	IV	117.4	1972	IV	177.1
1963	I	118.0			
1963	II	119.0			
1963	III	120.9			
1963	IV	122.6			

Sources: United States Department of Commerce, *Business Statistics* and *Survey of Current Business*, Washington, D.C.: United States Government Printing Office.

Table A-4
United States Crude Steel Capacity (Millions of Tons)

Year	*Quantity*	*Year*	*Quantity*
1954	124.3	1958	140.7
1955	125.8	1959	147.6
1956	128.4	1960	148.6
1957	133.5		

Source: American Iron and Steel Institute, as reported in Joint Economic Committee *Steel Prices*, p. 186.

Table A-5
United States Crude Steel Capacity Estimates (Millions of Tons)

Year	*ECE*	*Wall St. Jnl.*	*OBE*	*CEA*	*ECSC*	*Other*
1961	148.8	150	151	151.6		
1962	149.4	157	153	154.6		
1963	151.0	162	155	157.7		
1964	153.8	165	157	160.8		
1965	156.5	168	159	164.0	162.8	
1966		174	161	167.3		
1967		184	163	170.6		155
1968			165	174.0		
1969			168	177.5		
1970			171	181.0	178.2	180

Sources:

ECE: United Nations Economic Commission for Europe, *Steel Market Review*, New York: United Nations, various years.

Wall Street Journal: As reported in *Senate Finance Committee, Steel Imports*, p. 18.

OBE: Based on Office of Business Economics estimate of 1 ½% growth per year in Joint Economic Committee *Steel Prices*, p. 186.

CEA: Council of Economic Advisers estimate based on 2% growth per year in Council of Economic Advisors *Report to the President on Steel Prices.*

ECSC: European Coal and Steel Community estimates in Senate Finance Committee: in *Steel Imports.*

Other: 1967 Iron Age estimate in Senate Finance Committee, *Steel Imports*, 1970 Bethlehem estimate from Senate Judiciary Committee, *Administered Prices: Steel.* p. 604.

Table A-6
Capacity Utilization in the United States Steel Industry (Millions of Tons)

Year	*Quarter*	*Production*	*Capacity*	*Production / Capacity*
1954	I	22.32	31.43	0.710
1954	II	21.80		0.694
1954	III	20.10		0.639
1954	IV	24.05		0.765
1955	I	27.31	31.45	0.858
1955	II	29.89		0.950
1955	III	28.57		0.909
1955	IV	31.25		0.994
1956	I	31.24	32.10	0.973
1956	II	30.74		0.958
1956	III	20.16		0.628
1956	IV	32.44		1.011
1957	I	31.57	33.37	0.947
1957	II	28.98		0.869
1957	III	27.12		0.813
1957	IV	25.01		0.749
1958	I	18.79	35.17	0.534
1958	II	18.96		0.539
1958	III	21.31		0.606
1958	IV	26.12		0.743
1959	I	30.48	36.90	0.826
1959	II	33.79		0.916
1959	III	8.20		0.222
1959	IV	20.96		0.568
1960	I	34.74	37.15	0.835
1960	II	26.01		0.700
1960	III	19.64		0.529
1960	IV	18.88		0.508
1961	I	19.74	37.90	0.521
1961	II	25.11		0.663
1961	III	25.66		0.677
1961	IV	27.48		0.725
1962	I	30.63	38.65	0.793
1962	II	23.46		0.607
1962	III	20.52		0.531
1962	IV	23.46		0.613
1963	I	26.69	39.42	0.677
1963	II	32.55		0.826
1963	III	24.29		0.616
1963	IV	25.72		0.652

Table A-6 continued

Year	*Quarter*	*Production*	*Capacity*	*Production / Capacity*
1964	I	29.50	40.20	0.735
1964	II	31.80		0.791
1964	III	31.29		0.778
1964	IV	34.47		0.857
1965	I	35.04	41.00	0.855
1965	II	35.57		0.868
1965	III	32.82		0.800
1965	IV	27.74		0.677
1966	I	32.90	41.82	0.787
1966	II	35.16		0.841
1966	III	33.16		0.793
1966	IV	32.83		0.785
1967	I	31.63	42.65	0.742
1967	II	30.50		0.715
1967	III	30.35		0.712
1967	IV	34.42		0.807
1968	I	36.53	43.50	0.840
1968	II	37.05		0.852
1968	III	28.49		0.655
1968	IV	27.98		0.643

Sources:

Production: "Annual Statistical Review," *The Iron Age* (January 2, 1969), p. 128.

Capacity: Quarterly rates of capacity estimates made annually by the AISI and Council of Economic Advisers in Tables A-4 and A-5.

Table A-7
Collective Bargaining in the United States Steel Industry (1954-68)

Year	*Negotiations Begin*	*Strike Deadline*	*Settlement Date*
1954	May 18	June 30	June 29
1955	June 7	June 30	July 1
1956	May 28	June 30	August 3
1959	May 5	June 30	January 4[a]
1962	February [b]	June 30	March 31
1963	March 15	July 31	June 20
1965	March 28	May 1[c]	September 6
1968	May 5	August 1	July 20

[a]January 4, 1960.

[b]Negotiations began secretly.

[c]Strike deadline extended to September 6.

Sources:

United States Department of Labor, BLS, "Wage Chronology: United States Steel Corporation: 1937-1967," *BLS Bulletin*, no. 1603 (Washington, D.C.: United States Government Printing Office, July 1968), pp. 3, 17, 23, 24, 30, 31, 42, 43.

The Iron Age (July 4, 1968), p. 81.

Table A-8
Quarterly Periods of Steel Buying Affected by Steel Labor Negotiations

Year	*Quarter*	*Periods Affected*	*Imports (Millions of Tons)*	*Production*
1954	I	0	0.207	22.324
1954	II	1	0.324	21.807
1954	III	0	0.308	20.102
1954	IV	0	0.254	24.058
1955	I	0	0.253	27.316
1955	II	1	0.319	29.890
1955	III	1	0.364	28.578
1955	IV	0	0.427	31.252
1956	I	0	0.342	31.242
1956	II	1	0.427	30.748
1956	III	1	0.455	20.167
1956	IV	1	0.591	32.441
1957	I	0	0.433	31.595
1957	II	0	0.367	28.988
1957	III	0	0.259	27.120
1957	IV	0	0.249	25.011
1958	I	0	0.269	18.791
1958	II	0	0.386	18.962
1958	III	0	0.536	21.317
1958	IV	0	0.644	26.120
1959	I	1	0.791	30.488
1959	II	1	1.205	33.790
1959	III	1	1.191	8.201
1959	IV	1	1.436	20.961
1960	I	1	1.470	34.741
1960	II	0	0.874	26.013
1960	III	0	0.618	19.647
1960	IV	0	0.607	18.880
1961	I	0	0.461	19.741
1961	II	0	0.714	25.117
1961	III	0	0.805	25.668
1961	IV	0	0.917	27.488
1962	I	0	0.895	30.635
1962	II	1	1.046	23.463
1962	III	0	0.994	20.523
1962	IV	0	0.992	23.707
1963	I	0	0.888	26.593
1963	II	1	1.312	32.550
1963	III	0	1.531	24.293
1963	IV	0	1.403	25.725

Table A-8 continued

Year	*Quarter*	*Periods Affected*	*Imports (Millions of Tons)*	*Production*
1964	I	0	1.304	29.508
1964	II	0	1.565	31.806
1964	III	0	1.522	31.290
1964	IV	0	1.735	34.472
1965	I	1	1.747	35.043
1965	II	1	3.015	35.572
1965	III	1	2.868	32.824
1965	IV	0	2.563	27.745
1966	I	0	1.851	32.909
1966	II	0	2.559	35.164
1966	III	0	3.207	33.168
1966	IV	0	2.824	32.831
1967	I	0	2.367	31.637
1967	II	0	2.783	30.502
1967	III	0	2.869	30.358
1967	IV	0	3.285	34.423
1968	I	1	3.340	36.530
1968	II	1	4.709	37.056
1968	III	1	5.284	28.494
1968	IV	0	4.408	27.982

Sources: Periods Affected: See text.
Imports: See Table A/1.
Production: See Table A/6.

Table A-9
Man-Days Lost Due to Work Stoppages in the Water Transportation Industry (As a Percentage of Total Man-Days per Quarter)

Year	*Quarter*	*Man-Days*
1954	I	.084
	IV	.038
1955	I	.012
	II	.025
	III	.046
1957	I	.034
1959	II	.040
1961	I	.002
1962	III	.021
1963	I	.109
1964	III	.005
1965	I	.291

Source: See text.

Table A-10
The Share of Imports in the Domestic and International Market[a]

	Consumption / Imports	*Production / Imports*	*Imports / World Imports*
Belgium	2.6	9.1	.051
Canada	8.7	8.3	.038
France	4.0	4.2	.166
West Germany	6.0	7.2	.182
Italy	5.9	5.5	.103
Japan	138.7	161.9	.013
Luxembourg	2.6	9.1	.004
Netherlands	1.5	1.5	.080
United Kingdom	15.4	16.9	.051
United States	14.8	13.3	.309

[a]Ratios of quantities expressed in metric tons.

Sources: OECD, *The Iron and Steel Industry*, 1964, 1966, 1968.
United Nations Statistical Office, *Monthly Bulletin of Statistics*, (New York: United Nations, June 1968), p. 54.

Table A-11
The Share of Exports in Domestic and International Markets[a] (1967)

	Production / Exports	*Consumption / Exports*	*Exports / World Exports*	*World Exports / Imports*	*Rest of World Exports / Exports*
Belgium	1.2	0.2	0.164	45.48	5.10
Canada	13.9	12.9	0.013	44.58	NA
France	3.3	2.3	0.125	10.19	7.69
Germany	3.3	2.3	0.237	9.32	NA
Italy	8.7	7.7	0.038	16.45	25.03
Japan	6.7	5.7	1.950	123.53	4.40
Luxembourg	1.2	0.2	0.082	85.00	11.19
Netherlands	1.6	0.5	0.046	21.20	20.67
United Kingdom	6.7	5.7	0.075	33.10	12.20
United States	116.8	108.6	0.022	5.54	NA[b]

[a]Ratios of quantities in terms of metric tons.

[b]NA: not applicable.

Source: Same as Table A-10.

Index

abandonment, 145
accelerated depreciated, 152, 153
Adams, W., 12, 14, 23, 24
Adelman, M., 118, 130
administrative fees, 144
advertising rules, 151
Aliber, R., 19, 25, 118, 130
American Importers Association, 148
American Iron and Steel Institute (AISI), 5-7, 11, 16-18, 22-24, 27, 31, 32, 35, 47, 55, 56, 69, 162-164, 184, 186
American selling price, 142
American Society of Mechanical Engineers, 161
Annual Survey of Manufactures, 14
antidumping laws, 17, 123-127
Anti-Friction Bearing Manufacturers Association, 126
Ault, D., 13, 14, 23
Austria, 12, 14

Bailey, M., 54, 68
Bain, J., 15, 24
Baker, L., 149
balance of payments, 4, 18
Balassa, B., 31-35, 49, 70, 95, 118, 124, 128, 130, 132
Baldwin, R., 19, 20, 25
Ball, R., 52, 67
Basevi, G., 31, 32, 35, 36, 49
Bauge, K., 128
Belgian Office of Foreign Trade, 131, 163
Belgium, 6, 8, 14-16, 38-43, 61, 62, 64-66, 74-85, 87-100, 110, 112, 113, 117, 119-121, 124, 136, 138, 145, 152-154, 156-159, 165, 167, 171-175, 190, 191
Bell, J., 162
biased procurement, 73-78, 109-112
Bidwell, P., 19, 25, 148
Boner, J., 24
Bragg, G., 130
Btitish Iron and Steel Federation, 162
British Steel Corporation, 167
Brussels Definition, 140-146
Brussels Tariff Nomenclature, 138
Bureau of Labor Statistics (BLS), see United States
Burke-Hartke Bill, 17
Buy-American policy, 111, 129

Canada, 3, 4, 6, 8, 12, 14-16, 28, 37-43, 61, 62, 64-66, 73-78, 80-85, 88-94, 96-100, 109-112, 114, 117, 119, 122-126, 136, 138-142, 145, 146, 153, 154, 156-158, 161, 167, 190, 191,
Ministry of National Revenue, 141
Canadian-American Committee, 126
Caves, R., 176
Chang, T., 70
Cheng, H., 70
Chief component by value standard, 139
Chow, G., 69
classification rules, 138-140
coal, 7, 8, 155-157
Cohen, J., 23, 162
Common Market, see European Economic Community
consulting engineering service, 114
Consumers Union, 18
continuous casting, 13, 14
constructed value, 142
contract rate, 168
Convention on the Valuation of Goods for Customs Purposes, see Brussels Definition
Corlett, W., 70, 71
Cort, S., 23, 24
costs, 4, 9; labor, 4; capital, 9, 15; materials, 7, 8, 59; transportation, 8
Cotton, C., 49
Council of Economic Advisers (CEA), 14, 24, 56, 69, 169, 176, 184, 186
countervailing duty, 124
credit controls, 113
credit reference service, 114
Customs Cooperation council, 138, 147
customs: court, 141, 143; fees, 144; stamp tax, 144; warehouse, 145

de Hass, J., 19, 25
depreciation, 113, 152, 153
de Vries, B., 61, 71
Dillon Round, 28, 31, 35, 66
Dirlam, J., 12, 14, 23, 24
document requirements, 108, 136-138
Dosser, D., 130, 131
Drapkin, M., 70
Dresser, J. van B., 23, 162, 163
dual rate contract, 168
Due, J., 130
dumping, 123

Economic Commission for Europe, 6, 23, 29, 70, 184
elasticities: domestic demand, 58-60; domestic supply, 47, 82; export demand, 48, 66, 67; export supply, 48, 58-60, 63-66;

import demand, 44-46, 51, 58, 60-62; import supply, 44-46; income, 58, 79
Entente de Bruxelles, 165
escape clause, 17
Europe, 3, 8, 12, 28, 31, 116, 119, 146, 155, 167, 169
European Coal and Steel Community (ECSC), 56, 109, 156, 159
European Economic Community (EEC) 4, 9, 10, 15, 28, 29, 31-34, 62, 74, 94, 95, 101, 109, 110, 119, 122, 124, 126, 135, 138, 140, 146, 153, 156, 160
European Free Trade Area (EFTA), 74, 160
exchange control, 106
exclusive supply agreements, 167
export: information, 114; limitations, 17, 79, 108, 109; value, 142

Federal Maritime Commission, see United States
Ferguson, C., 70
Fetherolf, A., 162
final list, 142
Finsider Group, 165
Flueck, J., 54, 68
Foreign Trade and Investment Commission, 17
France, 4, 6, 8, 14-16, 37-43, 61, 62, 64-66, 74-85, 87-100, 106, 108-110, 112-114, 117-119, 125, 136, 138, 140, 144-146, 153-159, 165, 167, 171-175, 190, 191
freight rate discrimination, 19, 168-175
Friedman, M., 59, 70

Gantt, P., 128
General Agreement on Tariffs and Trade (GATT), 27, 114-116, 119, 120, 124-126, 129-133, 148, 176
Germany, West, 7-9, 12-15, 37-43, 61, 62, 64-66, 74-85, 88-94, 96-100, 109, 110, 117-119, 125, 136, 138, 140, 145, 146, 153, 154, 156-159, 165, 167, 171-175, 190, 191
Gott, E., 11, 23, 24
Greece, 14
Gross National Product (GNP), 55, 183
Grubel, H., 19, 25
Grunzel, J., 19, 25

Haight, F., 19, 25
Hall, C., 118, 130
Haller, H., 130
Harberger, A., 130
health rules, 151
hedge buying, 56, 57, 188, 189
Hemmindinger, N., 127-129
Heuser, H., 19, 25, 176
Hogan, W., 23, 128
House Ways and Means Committee, see United States
Hout, T., 23
Houthakker, H., 61, 71, 101

import licenses, 105
incidence: taxes, 116, 118; tariffs, 44; subsidies, 47, 48, 151, 153
Inner Six, see European Economic Community
International Chamber of Commerce 147
International Monetary Fund, 127, 148
investment incentives, 153
invoices, 136, 137, 140, 141
iron ore, 7, 8
Italy, 6, 14, 15, 43, 61, 62, 64-66, 74-78, 80-100, 108, 110, 113, 117, 119, 120, 122, 125, 138, 140, 144, 146, 153, 154, 156-159, 165, 167, 171-175, 190, 191

Jackson, W., 107
Japan, 4, 7, 9, 10, 12, 14, 16, 28, 31-34, 37-43, 61, 62, 64-66, 74-78, 80-100, 106, 107, 109, 110, 113, 114, 116-119, 125, 138, 140, 144, 145, 151, 153-158, 165, 167, 169, 170-175, 190, 191
Jessop Steel Company, 126
Johnson, H., 18, 19, 25, 30, 42, 48, 49, 130
Junz, H., 130

Kaplan, A., 176
Kelly, W., 19, 25, 27, 47, 101, 128, 129, 133, 147, 148, 163
Kemp, M., 62, 71, 132
Kennedy Round, 27, 28, 35, 73, 94, 95, 124, 126
Kindahl, J., 55, 69
Kindleberger, C., 69
Knapp, L., 128
Kravis, I., 53, 54, 68
Kreinin, M., 52, 61, 67, 70, 71
Kryzyzniak, M., 118, 130

labor disputes, 16, 56, 57, 187-190
lags, 52
Lanzillotti, R., 176
Latrobe Steel Company, 126
Lawrence, R., 24
Leither, J., 48
Lipsey, R., 53, 54, 68
Lister, L., 23, 24, 163, 176
Lorraine, 8
low-cost loans, 154
Lowery, B., 130
loyalty rebates, 152
Luxembourg, 6, 7, 15, 37-43, 61, 62, 64-66, 74-85, 87-100, 110, 112, 117, 119, 120, 122,

124, 136, 138, 145, 153, 154, 156-159, 165, 167, 171-175, 190, 191

MacAvoy, P., 176
Magee, S., 61, 71, 101
majoration, 158
Malabre, A., 24
Mansfield, E., 11, 23
market: research, 113, 114; other assistance, 113
marking regulations, 135, 136
Marwah, K., 52, 67
Massel, M., 19, 25, 128, 132, 162
Masson, F., 127-129, 132, 133, 146-148, 164
Maxwell Stamp Associates, 28
McAlister, H., 55, 68
McLouth Steel, 12
Melvin, J., 28
Mieszkowski, P., 130
Ministry for International Trade and Industry, 107, 114, 154
Molz, F., 24
Morgan, D., 70, 71
Musgrave, R., 118, 130

National Bureau of Economic Research (NBER, NB), 53-55
national security rules, 161, 162
Netherlands, 6, 7, 12, 13, 15, 16, 38-43, 61, 62, 64-66, 74-78, 80-85, 87-100, 110, 117, 119-121, 124, 138, 146, 153, 154, 156-159, 165, 167, 171-175, 190, 191
nontariff barriers: definition, 19, 20, 21; distribution, 37-41, 94; effective rates, 29, 42, 95, 100; nominal rates, 36, 38, 43; restrictive effects, 87-101
normal price, 140-146

Office of Business Economics, see United States
Organization for Economic Cooperation and Development (OECD), 6, 23, 24, 61, 73, 121, 128-131, 148, 162, 163, 177
Organization for European Economic Cooperation (OEEC), 14
Orr, D., 176
oxygen process, 12

paratariff barriers, 138
patents, 160
penalties, 83, 114, 115
Pincers, A., 130
Polak, J., 70
Portugal, 14
preferential interest rates, 113, 154
price controls, 159, 160
program contract, 155
public bidding, 110
purchases, overpriced, 109-112, 155

quotas, 17, 18, 19, 46, 78, 79, 108

Rapp, W., 23
rebates of taxes, 114-123, 155
research and development (R&D), 11, 13
restrictive practices, 166, 167
Rhemus, C., 24
Rippe, R., 16, 24, 51, 67
Roogovens, 15
Roth, W., 28
Ruhr, 15
rule of similitude, 139

Saar, 7, 15
salvage valuc, 153
Scott, M., 61, 71
Seifer, 13
Shorter, F., 71
similar goods rule, 141
Spain, 14
Speck, W., 128
state trading, 19
statistical tax, 144
steel: capacity, 10, 11, 55, 56, 184-186; cartels, 165; concentration, 15, 16; consumption, 16, 190, 191; economies of scale, 13, 15; exports, 3, 5, 16, 191; imports, 3, 5, 8, 10, 11, 16, 52, 179, 190, 191; integration, 7, 8, 14; prices, 8, 10, 11, 14, 15, 52, 53, 181, 182; pricing policies, 14; production, 3, 5, 16, 185, 186, 190, 191; productivity, 6, 7, 11, 12; wages, 6
Stein, H., 19, 25, 118, 130
Stigler, G., 55, 69, 176
Sterling transfer form, 108
straight line depreciation, 152
subsidies: direct, 112, 113, 151, 152; domestic, 47, 82, 83, 151-157; export, 47, 83, 112, 113; surcharges, 17, 18, 86, 122, 123
Sweden, 31, 32, 33, 34

tariffs: 17, 18, 19; effective, 29-36, 95, 100; nominal, 27, 31, 32, 89, 94, 95; rationale, 18; restrictive effects, 44-46; 94-97; unweighted, 27-29; weighted, 28, 29
taxes: 19; border, 158; cascade, 115, 118-122; direct, 115, 116; indirect, 115, 116, 157-159; internal, 155; occulte, 115, 119, 120, 122; plant, equipment, and power, 120; single-stage, 115, 119, 122; value added, 115, 119, 122
technical specifications, 160, 161
technology, 4, 10, 11

Temporary National Economic Committee, 70
Thorn, R., 4, 8, 9, 14, 23, 177
tied foreign aid, 114
Tin Plate Club, 167
Topping, F., 49
Towle, L., 49, 128, 129, 132, 148
trade fairs, 114
trademarks, 151

United Kingdom, 4, 6, 7, 9, 14, 15, 28, 29, 32-34, 37-43, 61, 62, 64-66, 74, 78, 80-85, 87-100, 108, 109, 110, 113, 117-119, 122, 125, 126, 137, 138, 140, 144, 146, 152-158, 160, 167, 171-175, 190, 191, National Coal Board, 156
United Nations, 174, Conference on Trade and Development, 20
United States: 3, 4, 6, 7, 10, 12-17, 28, 29, 31-36, 38-43, 56, 58-62, 64-66, 74-101, 106-112, 116-119, 125, 126, 135-139, 141-146, 152-154, 156-158, 159, 161, 165, 167, 168, 170-175, 179-191
  Agency for International Development, 73-78, 111
  Bureau of Labor Statistics (BLS), 7, 12, 53-55, 57, 68, 69, 187
  Census Bureau, 32, 136
  Commerce Department, 3, 21, 49, 52, 66, 68, 69, 71, 101, 102, 121, 127, 148, 149, 176, 180, 183
  Commission on Foreign Economic Policy, 147, 148
  Customs Simplification Act, 142
  Defense Department, 111, 112
  Federal Maritime Commission, 168, 176
  House Ways and Means Committee, 127-129, 131, 132, 147, 148, 162-164, 176
  Internal Revenue Service, 153
  Joint Economic Committee, 19, 25, 48, 68, 127-130, 132, 153, 154, 162, 171, 172, 176, 177, 184
  Office of Business Economics, 56
  Office of Civil and Defense Mobilization, 161
  Secretary of Transportation, 137
  Senate Finance Committee, 23, 24, 28, 31, 48, 71, 101, 129-131, 148, 153, 154, 162-164, 176, 184
  Senate Judiciary Committee, 24, 69
  Shipping Act, 168
  State Department, 18
  Tariff Act, 139
  Tariff Commission, 133, 146, 147, 161
  Trade Agreements Extension Act, 161
  Treasury Department, 127, 146-149
United States Steel Corporation, 7, 54

valuation procedures, 27, 28, 140-143
van Cleve, H., 128
Viner, J., 19, 123, 132
von Haberler, G., 132

Weidenhammer, R., 9
Whitely, J., 127-129, 132, 133, 146-148, 164
Whittlesey, C., 19, 25
Wholesale Price Index, 53-55
Wilkinson, B., 28

Yntema, T., 58, 70

About the Author

Craig R. MacPhee is associate professor of economics at the University of Nebraska, Lincoln, where he teaches economic theory, international trade, and industrial organization. He received the B.A. from the University of Idaho and the M.A. and the Ph.D. from Michigan State University. Dr. MacPhee has authored a study guide for economics; *Economic Organization in Medical Equipment and Supply* with R. D. Peterson of Colorado State University (Lexington Books, 1973); papers presented at the Midwest Economics Association; and a number of articles.